CLIMATE CULTURE CHANGE

Inuit and Western Dialogues with a Warming North

TIMOTHY B. LEDUC

UNIVERSITY OF OTTAWA PRESS
2010

The University of Ottawa Press acknowledges with gratitude the support extended to its publishing list by Heritage Canada through its Book Publishing Industry Development Program, by the Canada Council for the Arts, by the Canadian Federation for the Humanities and Social Sciences through its Aid to Scholarly Publications Program, by the Social Sciences and Humanities Research Council, and by the University of Ottawa. We also gratefully acknowledge the Faculty of Environmental Studies at York University whose financial support has contributed to the publication of this book.

Library and Archives Canada Cataloguing in Publication

Leduc, Timothy B., 1970-
   Climate, culture, change : Inuit and Western dialogues with a warming North / Timothy B. Leduc.

Includes bibliographical references and index.
ISBN 978-0-7766-0750-4

   1. Traditional ecological knowledge—Canada, Northern. 2. Inuit—Ethnobiology—Canada, Northern. 3. Climatic changes—Social aspects—Canada, Northern. 4. Climatic changes—Political aspects—Canada, Northern. 5. Climatic changes—Environmental aspects—Canada, Northern. I. Title.

GN476.7.L43 2010   304.2089'970719   C2010-907906-X

Our understanding of climate goes beyond scientific reports into a wider realm of thinking. When I look at a snowless street in January I may see a natural weather variation, or I may see a human artefact caused by greenhouse gas emissions ... Unlike, say, the orbits of planets, the climate in the future actually does depend in part on what we think about it. For what we think will determine what we do.

Spencer R. Weart, climate physicist

Sila [is] a strong spirit, the upholder of the universe, of the weather, in fact all life on earth—so mighty that his speech to man comes not through ordinary words, but through storms, snowfall, rain showers, the sea, through all the forces that man fears, or through sunshine, calm seas or small, innocent children.

Najagneq, Inuit shaman

One of the increasing effects of a raised awareness of climate change is that we are becoming wise not only to our dependence on and shaping by past weather events but also to the effects of historic events on the shaping of relations between us.

Anne Primavesi, eco-theologian

Nature has the last word, and humanity's obligation is to reflect that wisdom.

Jaypeetee Arnakak, Inuit philosopher

CONTENTS

PREFACE AND ACKNOWLEDGEMENTS

The first decade of the twenty-first century has offered an abundance of scientific, political, economic, and intercultural debates on climate change that have highlighted for me the complex challenge we now face. From successive United Nations meetings in places like Kyoto, Montreal, and Copenhagen to ongoing public debates in the Canadian media and our living rooms, there has arisen a cacophony of hotly contested opinions about the nature of climate change, its relation to excessive fossil fuel use, and what an adequate response will entail. Early on I became disenchanted with the dominant concerns of defining the scientific physics of climate change, questioning the validity of the research and debating the political economic costs of mitigation and adaptation strategies. While these are all important issues that are discussed in this book, they seem to often marginalize a deeper inquiry into the climate-culture relations intimated in the opening quotes by climate physicist Spencer Weart, Inuit shaman Najagneq, eco-theologian Anne Primavesi, and Inuit philosopher Jaypeetee Arnakak. Their thoughts represent some of the many Western and Inuit voices that led me to ask some very different questions concerning the way we think about a responsive climatic reality, and, if we attempt such an inquiry, what the potential implications are for climate research and politics. These questions were not fully developed when I started writing *Climate, Culture, Change* but rather gradually evolved as I brought my Canadian culture and Western climate research into dialogue with various Inuit.

With today's northern warming being described by some climate researchers as a global "canary in the mine," it seemed highly relevant to the Canadian context of my concerns to engage Inuit in a discussion about the cultural relevance of today's changes. Beyond the climatic relevance of Inuit perspectives, this book also derived some of its original impulse from my short pre-academic life as a social worker in some of the indigenous communities of northern Labrador. It was here I first realized the extent to which my education in a Canadian and Western system of thought is contrasted, and often in conflict, with indigenous ways of living and thinking. This experience left me with questions about intercultural relations, colonial injustices, and the environment that continued to reverberate as I engaged various Inuit and Western perspectives on climate change. The Western eco-theology of Anne Primavesi proved extremely helpful as I contemplated the relation of northern warming to a history of social injustices.

An equally influential Inuit voice for this book and, more importantly, in the international climate change debate is Sheila Watt-Cloutier, the former chair of the Inuit Circumpolar Conference—now known as the Inuit Circumpolar Council—who was runner-up for the 2007 Nobel Peace Prize. In concert with many other Inuit, as well as Primavesi, her climate critique challenged me to consider the politics of Western nations like Canada, the economics of never-ending growth, and the nature of science itself. It is just such Canadian and, more broadly, Western cultural assumptions that the dialogues represented in this book connect with the contemporary change of northern *Sila*, or weather, and the global climate system.

Though the chapters that follow are very much grounded in the Canadian political and environmental realities of my experience, the global scale of climate change and the historic roots of Canadian-Inuit relations led me to consider the broader cultural influence of Western European, American, and international forces. The social context of Canadian environmental politics was highlighted for me by many Inuit who talked of the North's present climate instability in relation to fairly recent colonial changes. Such perspectives surrounded my writing with questions about the historic influence of various Western traditions on climate change, including Christian, economic, and political missions to the North, which, over the past 150 years, were increasingly informed by the emerging

Canadian nation. These challenge the common political and public view of climate change as a contemporary issue related solely to unsustainable energy practices. I approached this challenge by thinking about climate change's relation not only to a host of environmental issues but also to the social impacts of colonialism—all of which successively emerged in the North as Inuit-Canadian relations intensified over the twentieth century. Consequently, I have tried to contextualize my Canadian cultural, political economic, and academic position within both the arising global debate on what a sustainable response to a global climate system will entail and various aspects of Western cultural history.

Facilitating this intercultural dialogue between Inuit and Western climate thought has been an interesting experience that has allowed me to meet many insightful people whom I would like to acknowledge before moving on to the Introduction. Particularly influential were the many Inuit whose words ebb and flow over these pages, for without their thoughts, direction, insights, and advice this book would not have been possible. There are many individuals like Watt-Cloutier and Najagneq who I only know through their advocacy, writings, or earlier discussions with researchers, but who nonetheless deserve thanks for guiding my thoughts and writing. The same can be said of some important organizations like the Inuit Circumpolar Council and Inuit Tapiriit Kanatami. On a more personal level, I delivered a two-day climate change workshop in Chesterfield Inlet, Nunavut, which was attended by eight Inuit who were very helpful. I am extremely thankful for being introduced to Simionie Sammurtok, president of the local Hunter and Trapper Organization, who was the workshop co-facilitator and translator. His concern about climate change and support of the workshop were indispensable. Other participants in Chesterfield Inlet who deserve thanks include Andre Tautu, Eli Kimmaliardjuk, Louis Autut, Casemir Kriteedhuk, Elizabeth Tautu, Bernie Putulik, and Mark Amarok. The climate change, environmental, and social observations these individuals offered, based largely on their hunting experiences, were significant and inform each chapter. All the workshop materials, recorded tapings, and written work with Inuktitut translations, as well as my related publications, were sent to Simionie Sammurtok so they could be archived for the community as a symbol of the gratitude I feel toward those who voiced their concerns. I do hope these materials, in concert with this book, may be of relevance to Chesterfield Inlet, and other Inuit, as they continue responding to the North's changes.

My northern trek to Chesterfield Inlet was originally inspired by Jaypeetee Arnakak, an Inuit philosopher and policy worker based in Iqaluit, Nunavut, who I am forever indebted to. Over a four-year period, Arnakak corresponded regularly with me about various Inuit concepts and policy issues and their relation to the political and economic realities of my climate concerns. Our dialogues were so influential in helping me conceptualize the first and sixth chapters in the book that I asked him if he wanted to be recognized as a co-author for these. He humbly declined my offer, because he felt the chapters were ultimately my own representation of the dialogue and related research, and this, I agreed, was a valid point. With that in mind, I need to clarify here that I am no expert in Inuit culture or northern ways of living—one would at least have had to live in the North in order to make that claim. Rather, all I try to do here is truthfully and respectfully represent the words of Inuit such as Jaypeetee Arnakak, Simionie Sammurtok, Sheila Watt-Cloutier, and Najagneq, while also considering their potential implications for Canadian and Western ways of thinking about climate change. Any faults in representing and interpreting their thoughts are ultimately due to my own personal and cultural limitations as a researcher based in the more southerly urban reality of Toronto, Ontario.

This book has also been inspired by various Western ethnographers whose work with Inuit and other indigenous people helped define my intercultural research of climate change. The early-twentieth-century writings of Knud Rasmussen on his meetings with Inuit such as Najagneq, in places such as Chesterfield Inlet, offered me a wealth of knowledge that still informs current research in fields as diverse as anthropology, cross-cultural studies, resource management, and northern climate change. Other pivotal guides were anthropologists Julie Cruikshank, George Wenzel, and Mark Nuttall, whose writings significantly inform the pages that follow. Also particularly helpful has been the research of Fikret Berkes on Inuit and indigenous adaptation to northern environmental and climate change. I am indebted to his thoughts and, on a personal note, the help he provided me in securing a small research grant to conduct the Chesterfield Inlet workshop. In this regard, I must also thank the Integrated Management Node of the Ocean Management Research Network and Social Sciences and Humanities Research Council of Canada for funding the Chesterfield Inlet workshop, as well as the Nunavut Research Institute for licensing the research. By bringing such ethnographical research into dialogue with Inuit like Jaypeetee Arnakak and Sheila Watt-Cloutier, I have created a book that seeks to add critical intercultural dimensions to our sense of Canada's and the West's role in northern and global climate change.

The process of undertaking this dialogue has taught me much about Inuit culture and knowledge, but it has revealed even more about my own Canadian cultural heritage and why new ways of thinking about climate culture relations are needed. It is for this reason I began the book with the quote from climate physicist Spencer Weart on the relation of climate change to our thought. His insight arose from a history of Western climate research and politics that informs many of the chapters that follow. Perhaps most influential in helping me to consider the changing relation between Western environmental research, Canadian political economics, and climate thought was my primary dissertation advisor Ray Rogers. He was with me throughout this project, giving generously of his time in reading successive drafts and guiding me toward environmental critiques of Western culture and Canadian policy. Important insights, edits, and ideas were also provided during the dissertation-writing by Stephen Scharper, Mora Campbell, Peter Timmerman, and Heather Eaton, all of whom broadened the interdisciplinary scope of what I brought into dialogue with the Inuit.

On an institutional level, I am indebted to the Faculty of Environmental Studies at York University in Toronto for having a program with the interdisciplinary flexibility to support my research. It was the only program I applied to, and, looking back, I do not know of any other place in Canada where my research would have unfolded with such breadth of interdisciplinary and intercultural consideration. I am also thankful for the financial support I received from York University in the form of the President Dissertation Scholarship, as well as the Ontario Government for its graduate scholarships. Regarding the process of transforming my dissertation into a book, I need to thank Michael O'Hearn and Eric Nelson at the University of Ottawa Press. Their invaluable guidance, in concert with Stephen Scharper's advice and the editorial comments provided by the manuscript's two reviewers, were essential in helping me round out the intercultural and interdisciplinary thought on climate change found here.

To end, I want to extend my heartfelt thanks to family and friends who helped support this project in a variety of ways. At various points I received editorial advice and/or guidance from friends like Scott, Rebecca, and Lorena. Throughout all of the stages of writing, Mark provided great editorial advice and then, at the end, offered some critical insights that came to shape the Introduction and Conclusion. Others, like Neil, Jude, and Dufferin, provided an ever-supportive presence while I wrote. Ultimately, the biggest support was given by my partner Christina. From the dissertation's first drafts to the book's final edits, she tirelessly offered her thoughts, edits, advice, and loving support. Finally, I thank our daughter, Iona, whose arrival during my writing of this book made Najagneq's seeming obscure relating of "innocent children" with *Sila* and, on a broader scale, climate, make sense. For if the thesis of *Climate, Culture, Change* is sound, the future we leave our children will depend on our cultural capacity to change the ways we think about climate-human relations.

INTRODUCTION:
ENDANGERED KNOWLEDGE

This book is about climate change in the Canadian North, and the challenge it poses to both Western and indigenous knowledge. Many researchers have referred to the North as the climatic equivalent of the "canary in the mine," the place where the onset of climate change will first be felt before triggering changes in other regions of the world.[1] Yet such changes are affecting both physical ecologies and cultural knowledge systems, in terms of both the North's indigenous inhabitants and more southerly based researchers. The Inuit hunter Andre Tautu offered me insight into the scope of this challenge as it is being posed to indigenous people, explaining that, in the past, Inuit knowledge or *Inuit Qaujimatuqangit* (IQ) was real because things were stable, but the "weather has so changed that IQ is pretty much gone, it can no longer predict because of the change in climate."[2] As I listened to Inuit voices like that of Tautu, there seemed to be a parallel with the challenges currently facing a broad array of knowledge that is intimately related to the Western tradition. From the science of the Intergovernmental Panel on Climate Change (IPCC) to the politics of the Kyoto Protocol, a host of uncertainties are impinging upon Western nations, scientific communities, economic interests, and religious groups as they try to understand and respond to these changes. While the chapters that follow examine these challenges to both Inuit and Western knowledge alike in the Canadian context of my research, they are more broadly concerned with an intercultural dialogue that can potentially expand our communal thoughts and responses to this northern and global climate threat.

In January 2009 we had an opportunity to see some of the difficulties these different types of cultural knowledge have

1

interacting when Canadian Environment Minister Jim Prentice convened a Polar Bear Roundtable of governmental wildlife managers, environmental scientists, and Inuit representatives. The Conservative government had the preceding year released a national report on the "special concern" of the polar bear that highlighted the importance of conserving this wildlife icon that has both cultural significance for Canadians and "cultural, spiritual, and economic significance to some northern native peoples."[3] Unfortunately, the report's absence of any climate change analysis led researchers like Polar Bears International scientific advisor, Dr. Andrew Derocher, to argue that no credible polar bear scientist could support its conclusions. At the roundtable, Derocher's critique continued as he, in concert with other environmental organizations like World Wildlife Fund, expressed concern that the gathering largely ignored the climate change issue in favour of a conservation-based hunting focus, despite northern warming being "the main threat to polar bears."[4]

In between these conflicting Western views fell the Inuit Tapiriit Kanatami and the Inuit Circumpolar Council, two Inuit organizations that support the scientific call to deal with climate changes that are "being felt by Inuit communities," but differed politically from scientists on the current threat to polar bears.[5] As Inuit Circumpolar Council President Duane Smith put it, "[T]here is a need to reduce or avoid rhetoric, hype, and alarmist claims related to Polar Bear extinctions and over-hunting, including the condemnation of the sport hunt as anti-conservation."[6] This tension between the Inuit, environmental research, and Canadian politics has a long history that is important to consider in relation to today's polar bear and northern warming issues.

While the Inuit Circumpolar Council and Inuit Tapiriit Kanatami came to the roundtable arguing that Inuit wildlife management and IQ have allowed polar bear populations to double over the past three to four decades, the Polar Bears

International science of Derocher projected Canada's estimated twenty-two thousand polar bears—which constitute about 55 to 65 percent of the world population[7]—will plummet by two-thirds before "2050 as the ice retreats."[8] Such Inuit-environmentalist conflicts can be seen as a continuation of intercultural dynamics described by anthropologist George Wenzel in relation to the late-twentieth-century political debate on the seal hunt.[9] The dominant position of the animal rights movement at this time was "that Inuit culture has undergone such extensive technological change that it no longer resembles its traditional antecedents."[10] This is remarkably similar to the recent Polar Bears International assessment that Inuit claims of an increasing polar bear population lack validity because they have arisen from a nontraditional sport hunting industry that brings $30,000 per bear to communities with "almost no other source of outside income."[11] From such a perspective, the position of the Inuit Circumpolar Council and Inuit Tapiriit Kanatami can be dismissed not only because these organizations are economically and culturally biased, but, more broadly, because they recently evolved to defend Inuit interests from Western government, scientific, and industry interests. Unfortunately, such an argument also undermines the support environmental groups may receive from these same Inuit organizations when it comes to a common and interrelated concern like climate change.

During the seal hunt debate, Wenzel explains Inuit groups like Inuit Tapiriit Kanatami responded to similar environmentalist positions by arguing such a view continues colonial trends that have dissociated Inuit from their traditional northern relations and IQ. In other words, Inuit culture has been forced into a false dichotomy of either becoming more interconnected to Canadian and international forces and thus lose the claim of being indigenous, or of practicing a culture that is not allowed to evolve despite the ecological impact of Western political economic forces. A later analysis of Wenzel poignantly connects the rise of polar bear sport hunting in the early 1980s to both the economic impact of the seal hunt ban and government initiatives to shift northern economic development.[12] Despite this new cultural approach being "a highly non-traditional use of polar bear," Wenzel points out this does not mean the traditional value of polar bears to Inuit culture and IQ has been forgotten.[13] In other words, the Inuit position represented by these groups is informed by an IQ that is not a relic of the past, but rather dynamically responsive to changing ecological, climatic, and intercultural relations.

Particularly influential in helping Wenzel to consider the multiple dimensions of IQ was Jaypeetee Arnakak, an Inuit policy analyst and writer who corresponded personally with Wenzel and publicly through the northern media. In his analysis, Wenzel cites a newspaper article by Arnakak that defines IQ as "a 'living technology' through which Inuit 'thoughts and actions,' 'tasks and resources,' 'family and society' are organized."[14] Reflecting upon this knowledge's historic depth, a recent Inuit compilation further describe it as being "passed on to us by our ancestors, things that we have always known, things crucial to our survival, patience and resourcefulness."[15] It is the long relation of IQ to this northern ecology that makes it particularly helpful for climate research. As adaptive management researchers Dyanna Riedlinger and Fikret Berkes state, Inuit ancestral knowledge provides a way for dealing with the "lack of historical baseline data against which to measure data" on northern ecological processes.[16] This argument for complementary knowledge had some success at the roundtable, for while environmental researchers raised concerns that the climatic dimensions of the polar bear issue were being neglected, both Inuit and the Conservative government committed to integrating "Inuit traditional knowledge and science to build a better understanding about the changing environment and polar bear."[17] Essential to such an intercultural initiative would be, according to Duane Smith of the Inuit Circumpolar Council, the fostering of respect between Inuit, scientists, and government officials.[18]

Despite this potential for complementing Western research and IQ, both Wenzel and Arnakak have expressed concern about how these different types of cultural knowledge are being brought together. Perhaps the most common critique is the equating of IQ and other indigenous kinds of knowledge with what researchers define as traditional ecological knowledge. While Wenzel explains such a reduced view of IQ "diminishes the depth of its socio-cultural content,"[19] both Fikret Berkes and anthropologist Julie Cruikshank propose such a framing maintains an inequality between indigenous and Western knowledge.[20] Based on her northern indigenous research, Cruikshank argues traditional ecological knowledge recasts earlier Western ideas like "primitive superstition, savage nobility, or ancestral wisdom" in such a way that local indigenous knowledge continues to be engaged "as an object for science rather than as a kind of knowledge that could inform science."[21] From an Inuit perspective, Arnakak explains that IQ is neither ecological knowledge for managing the environment nor traditional knowledge isolated in the past. Talking about the latter point, he writes:

> The fact remains that *Inuit Qaujimajatuqangit* is a semi-literal translation of the original term in English—and in the passive tense at that. I have suggested on a number of occasions taking out the reference to "old" in *Qaujimajatuqangit*, and making the term an infinitive—Inuit *Qaujimaningit*—or simply, Inuit Knowledge.[22]

According to Wenzel, the reason IQ's traditional ecological knowledge most commonly enters Western environmental interpretations is because it "is that which is most comprehensible."[23] Consequently, those "spiritual or ideological" aspects of IQ that Inuit use to interpret traditional ecological knowledge are often marginalized in a nonempirical or, in Arnakak's terms, traditional category that is "viewed as possessing lesser substance."[24] The result for Western researchers is a missed opportunity to engage indigenous thought that could perhaps help us interculturally think about a climate system that is interconnected to human practices.

In response to these critiques, I have here attempted to follow a post-colonial approach that moves "away from expert-knows-best science and towards accepting indigenous knowledge as a source of knowledge that complements science."[25] Though my background as an environmental researcher means this book is primarily informed by interdisciplinary analyses rather than an ethnographic exploration of northern warming, the need to ensure that I was not objectifying and universalizing IQ required me to follow Wenzel, Cruikshank, and Berkes into a dialogue with Inuit. Particularly important in this regard were four years of correspondence on IQ and climate with Jaypeetee Arnakak, the same Inuit philosopher and policy analyst who informed Wenzel's research and whom I met in 2003 after he lectured at an academic "Sovereignty, Indigenous Knowledges and Environment" speaker series in Toronto. In contrast to those helpful ethnographies that detail the political economic, ethical, and spiritual relations northern indigenous people have with an ecology that is based on a more sustained relation, I tried to grasp IQ's multiple dimensions by bringing an interdisciplinary analysis of ethnographic, resource management, and climate research into dialogue with diverse—sometimes conflicting—Inuit voices like that of Arnakak, the Inuit Circumpolar Council, and others from the present and the past. Such an approach practices what qualitative researchers refer to as triangulation, the use of multiple methods "to secure an in-depth understanding" that has "rigor, breadth, complexity, richness, and depth."[26] The following chapters represent this dialogue in the form of various IQ concepts and stories that complement and deepen some of the more dominant Western and Canadian views on northern and global climate change.

An intimation of how IQ and climate research will interact in this book can be sensed in a story told by Wenzel of an Inuit hunt that returned with a polar bear. The hunting party left Clyde River, a hamlet on the northeast shore of Nunavut's Baffin Island, searching for seals during an uncertain time when "the lack of snow on the ice meant that hunters would probably be more easily detected by seals" and their other main predator, polar bears.[27] While on the three-day hunt, Wenzel and his unnamed Inuit companion sighted a den from which a polar bear emerged. Both men found it odd that the bear stood motionless and simply observed them, and they each formed their own cultural assessment of its behaviour. In contrast to Wenzel's conclusion that the community's need for food required a quick kill, the IQ of his northern guide slowed down the approach so as to study the bear's peculiar behaviour for almost three hours. The hunter informed Wenzel that "he had never seen a bear acting as this one did, that he did not understand the bear and that I should be patient." Only after they came within twenty-five metres of the bear, the animal observing them the whole time, did the hunter shoot and kill his prey. In the village the details of the hunt were recounted, with the hunter repeatedly referring to "the odd behaviour of the bear and how he had become convinced that the successful conclusion of this hunt would turn on discerning how the animal wanted to be approached." This experience clarified for Wenzel that "the hunter not only acts with respect towards the animal, but does so in a way that he has discerned from the actions of the bear."[28] The implication I took from this story was that the way in which IQ's ethical and spiritual dimensions contextualize traditional ecological knowledge may have some important insights if considered in relation to the "odd behaviour" of today's polar bears and climate changes.

Over the past decade, such Inuit stories of polar bear hunts have been joined by an emerging story of northern warming and increasing polar bear uncertainty. My first contact with this story occurred while talking with a group of elders and hunters about northern climate change from the Nunavut community of Chesterfield Inlet on the northwest coast of Hudson Bay. These voices, which were introduced in the Preface and Acknowledgements, came to inform my sense of IQ through Arnakak's suggestion that I engage an Inuit community where first-hand accounts can deepen both our conversations and other research findings. While Arnakak could offer me a sense of IQ's philosophy, and its relation to Western concepts and potential policy implications, he stressed the importance of talking with Inuit who were hunting on the land and intimately experiencing the effects of climate change and shifting polar bear behaviour. In October 2004 I facilitated a two-day workshop that brought into dialogue IQ and Western research. During one exchange, Andre Tautu spoke of staying up all night at his camp and leaning up against a door to keep out a hungry polar bear. Echoing the experience of the other participants, he compared his childhood memory of only seeing three or four bears a year with the present experience that sees them "crawling all over the country."

As with the assessment of the Inuit Circumpolar Council, Tautu held that scientists who view the polar bear as disappearing are wrong because their constant presence is making it difficult for Inuit like him "to make a living at commercial fishing." The Western climate research he and many other Inuit are critical of is epitomized in the reports of the IPCC that state the western Hudson Bay population of polar bears have declined "from 1,200 bears in 1987 to fewer than 950 in 2004" due to the seasonal decrease in sea ice from which they hunt.[29] Such decreases are not being confirmed by the experiences of Tautu, and this position is not confined to Chesterfield Inlet. As Arnakak explained to me, hunters throughout Nunavut are having "more incidents of problem bears, and more sightings where they rarely occurred."[30] This issue was highlighted a month prior to the roundtable in a national news story on the impact of hungry bears on Arviat, a community south of Chesterfield Inlet.[31]

With the Hudson Bay communities asked to refrain from hunting polar bears to help conserve their numbers, Arviat was inundated with bears wandering up the coast in search of food. As Polar Bears International scientist Andrew Derocher informed the press, the bears "want to get back on the ice in the fall to hunt for seals," but the unusually late freeze-up this year means they are ending up in coastal communities like Arviat.[32] These changes in human-bear interactions informed the Inuit Tapiriit Kanatami and Inuit Circumpolar Council position at the roundtable, with their media release stating that these increasing interactions "are posing safety and property damage concerns for Inuit."[33] Beyond Derocher's scientific description of the Arviat polar bear situation as one where "we're seeing little cracks in the ecology of the animal everywhere we look," it can be argued that these ecological cracks in combination with the international pressure to save the bear are creating an even more difficult cultural challenge for Inuit communities and IQ than the earlier seal debate.

While researching northern indigenous thought, Hugh Brody was taught that *irski* is the Inuit root word for the kind of fear Inuit from communities like Arviat and Chesterfield Inlet may experience today in relation to the increasing presence of polar bears. As he writes, "polar bears are sometimes said to be *irksina-*, terrifying."[34] Such an emotion makes sense for people who have had to negotiate a complex relation with a species that is sometimes their prey and at other times their predator. In a book on Alaska's northern warming, Charles Wohlforth describes an incident that clarifies why Inuit would conceive the polar bear as *irksina*.[35] As with Arviat, he writes that a plague of polar bears landed around the Alaskan community of Point

Lay during the early 1990s, which resulted in various initiatives aimed at protecting the people. Despite attempts to increase safety, the deep darkness of a December night saw Carl Stalker and his pregnant wife meet up with a hungry male bear as they walked home. While Carl protected his wife, she ran to get help. By the time a tracking party arrived on the scene, there was only a trail of blood that led out onto the sea ice to a bear "hunched over Carl's bones, which in two hours time had been stripped of flesh."[36] The climatic relevance of this story for Wohlforth was that bears are becoming more dangerous and unpredictable for human communities as they become less healthy due to the earlier loss of ice. It is clear that northern climate changes are not only impacting species like the bear but are also increasing IQ's uncertainty and the *irksina* experience of the Inuit.

This uncertain feeling is today not simply a function of changing polar bear behaviour; in Chapter 1 we will consider the way in which the North's warming weather or, in Inuktitut, *Sila*, is presenting a very dangerous challenge to both IQ and interdisciplinary climate research. A sense of how these changes are impacting Western knowledge can be gleaned from the analysis of climate physicist Spencer Weart. Despite the great strides made by interdisciplinary initiatives like the IPCC in increasing climate knowledge over the past few decades, he points out these endeavours have "scarcely narrowed the range of uncertainty."[37] Since the central factor behind this paradox is increasing greenhouse gas emissions that depend upon human actions, he argues the future climate "actually does depend in part on what we think about it."[38]

This assessment of Western climate knowledge, which is the focus of Chapter 2, suggests researcher assumptions are deeply intertwined with not only today's northern warming, but also with the future viability of other types of cultural knowledge like IQ. As we will see, such a critical perspective suggests Western climate research and environmentalism are no less informed by cultural biases than the claim laid against the Inuit

Tapiriit Kanatami and Inuit Circumpolar Council. It may be that a successful climate response will require a fundamental change in the way Western research engages other kinds of knowledge like IQ—a change that is no less fear-invoking in the context of today's global climate changes than Tautu's description of an IQ endangered by increasingly odd polar bear behaviour and northern warming.

After considering the threat northern warming and global climate change pose to IQ and Western knowledge, I shift the inquiry in Chapters 3–5 toward an analysis of Canadian, American, and Western political economic forces that are presently, as in the colonial past, eliciting a different kind of Inuit fear. Contrasting the *irksi* of a polar bear encounter is *ilira*, a fear described to Hugh Brody as being associated with "people or things that have power over you and can be neither controlled nor predicted."[39] The sources of *ilira* include "ghosts, domineering and unkind fathers, people who are strong but unreasonable, whites from the south," all of whom can make Inuit feel vulnerable. Explaining the specific role of Westerners in this fear, Brody writes that the history of Western-Inuit relations has been one of colonial powers who have acted very much "like that of ghosts—appearing from nowhere, seemingly supernatural and non-negotiable." The threat of these ghostly powers goes beyond the endangerment of Inuit culture and IQ, for a number of elders also complained to Brody that southern government and scientific officials seem to "care more about polar bears than about Inuit children."[40] They continued their critique by not only protesting that their hunters are being blamed for animal population declines, but added that it is in fact Westerners "who were doing damage to the land" and making Inuit feel *ilira*. By expanding upon research that indicates historic greenhouse gas emissions have predominantly arisen from developed nations like Canada, the United States, and the European Union,[41] I consider the extent to which Inuit are justified in connecting their *ilira* and *irksina* to the politics, economics, science, and religion of Western and, more specifically, Canadian culture.

These evolving critiques inform Chapter 6, the book's final chapter, as I return to one last Inuit view on how IQ and Western knowledge can complement each other's understandings of this northern warming that is radiating from the industrialized world to the south. Building upon the preceding elucidation of factors that are limiting such intercultural research, this chapter concludes by intertwining these two worldviews in a respectful dialogue on polar bears and other climatic animal changes that is grounded in an Inuit sensibility. It is important to point out here that I am not claiming either a universally accepted Inuit or Western worldview, but rather, as will be seen, I am following some dominant conceptual understandings that can be clarified through considering culturally internal and external differences of opinion. This intercultural attempt to transcend limiting assumptions leads me to an interesting proposal that we will examine over the following pages. Perhaps neither IQ nor Western knowledge is completely wrong or right in its view of today's northern and global changes, but rather their seemingly contradictory interpretations may derive from culturally defined understandings of ecology and climate that can be broadened and contextualized through intercultural dialogue. Thinking about the value of such an approach to today's climatic cultural changes is in fact the central thread that interconnects all of the chapters in *Climate, Culture, Change*. With that in mind, it is now time to follow the northward movements of the polar bears toward a spiritually and ethically expanded engagement of climate change and the North's warming *Sila*.

ENDNOTES

1   Mark Nuttall and Terry V. Callaghan, "Introduction," in Mark Nuttall and Terry V. Callaghan, eds., *The Arctic: Environment, People, Policy* (Amsterdam: Harwood Academic Publishers, 2000), xxv; also see O. Anisimov and B. Fitzharris, "Polar Regions (Arctic and Antarctic)," in contribution of Working Group II to the TAR of the IPCC, *Climate Change 2001: Impacts, Adaptation, and Vulnerability* (Cambridge: Cambridge University Press, 2001).

2   In October 2004 I conducted a two-day workshop with various Inuit hunters and elders from Chesterfield Inlet, Nunavut. They included Andre Tautu, Simionie Sammurtok, Eli Kimmaliardjuk, Louis Autut, Casemir Kriteedhuk, Elizabeth Tautu, Bernie Putulik, and Mark Amarok. This workshop is further introduced in this Introduction and in Chapter 1.

3   Committee on the Status of Endangered Wildlife in Canada (COSEWIC), *COSEWIC Assessment and Update Status Report on the Polar Bear in Canada* (Ottawa: Environment Canada, 2008).

4   Polar Bears International, "National Roundtable on Polar Bears, Media Release," 16 January 2009, http://www.polarbearsinternational.org/in-the-news/polar-bear-roundtable/.

5   Inuit Tapiriit Kanatami and Inuit Circumpolar Council, "Inuit of Canada Expect Substantial Consultations prior to Canadian Polar Bear Listing," Joint Media Release, 20 January 2009, http://www.itk.ca/media-centre/media-releases/inuit-canada-expect-substantial-consultations-prior-canadian-polar-bear-.

6   Ibid.

7   COSEWIC, *COSEWIC Assessment and Update Status Report.*

8   Margaret Munro, "Polar Bears Face Uncertain Future in Arctic," *Dose*, 28 December 2008, http://www.dose.ca.

9   George Wenzel, *Animal Rights, Human Rights: Ecology, Economy and Ideology in the Canadian Arctic* (London: Belhaven Press, 1991).

10   Ibid., 164.

11   Canadian Marine Environment Protection Society, *Chemistry, Calibre and Climate: The Plight of Canada's Polar Bear* (Vancouver: Canadian Marine Environment Protection Society, 2005), 5.

12   George Wenzel, "Polar Bear as Resource," (paper delivered at the Third Northern Research Forum [NRF] Open Meeting, Yellowknife, NT, 2004).

13   Ibid., 10.

14   George Wenzel, "From TEK to IQ: Inuit Qaujimajatuqangit and Inuit Cultural Ecology," *Arctic Anthropology* 41, no. 2 (2004): 242.

15   John Bennett and Susan Diana Mary Rowley, *Uqalurait: An Oral History of Nunavut* (Montreal: McGill-Queen's University Press, 2004), xxi.

16   D. Riedlinger and F. Berkes, "Contributions of Traditional Knowledge to Understanding Climate Change in the Canadian Arctic," *Polar Record* 37, no. 203 (2001): 315–328.

17   Environment Canada, "News Release: Minister Prentice Highlights Progress Made at Polar Bear Roundtable," 16 January 2009, http://www.ec.gc.ca/default.asp?lang=En&n=714D9AAE-1&news=24AABBD9-00C3-4E80-9517-2D37013C5FAF.

18   Inuit Tapiriit Kanatami and Inuit Circumpolar Council, "Inuit of Canada Expect Substantial Consultations."

19   Wenzel, "From TEK to IQ," 248.

20   Julie Cruikshank, "Uses and Abuses of 'Traditional Knowledge': Perspectives from the Yukon Territory," in D. G. Anderson and M. Nuttall, eds., *Cultivating Arctic Landscapes: Knowing and Managing Animals in the Circumpolar North* (New York: Berghahn Books, 2004).

21   Ibid., 21.

22   Jaypeetee Arnakak, *A Case for Inuit Qaujimanituqangit as a Philosophical Discourse* (Iqaluit, NU: JPT Consulting, 2004), 1.

23   Wenzel, "From TEK to IQ," 244.

24   Ibid.

25   Fikret Berkes, "Epilogue: Making Sense of Arctic Environmental Change?" in I. Krupnik and D. Jolly, eds., *The Earth Is Faster Now: Indigenous Observations of Arctic Environmental Change* (Fairbanks, AK: Arcus, 2002), 340; also see Fikret Berkes, *Sacred Ecology: Traditional Ecological Knowledge and Resource Management* (Philadelphia: Taylor & Francis, 1999).

26   Norman K. Denzin and Yvonna S. Lincoln, "Introduction: The Discipline and Practice of Qualitative Research," in N. K. Denzin and Y. S. Lincoln, eds., *The Sage Handbook of Qualitative Research* (Thousand Oaks, CA: Sage Publications, 2005), 5.

27   Story told in Wenzel, "From TEK to IQ," 246–247.

28   Ibid., 247.

29   IPCC, *Climate Change 2007: Climate Change Impacts, Adaptation, and Vulnerability, Summary for Policymakers, Contribution of Working Group II to the Fourth Assessment Report of the IPCC* (Geneva, Switzerland: IPCC Secretariat/ World Meteorological Organization, 2007), 231.

30   E-mail correspondence, 6 August 2004.

31   See Munro, "Polar Bears Face Uncertain Future in Arctic."

32   Ibid.

33   Inuit Tapiriit Kanatami and Inuit Circumpolar Council, "Inuit of Canada Expect Substantial Consultations."

34   Hugh Brody, *The Other Side of Eden: Hunters, Farmers and the Shaping of the World* (Vancouver: Douglas & McIntyre, 2000), 42.

35   Charles P. Wohlforth, *The Whale and the Supercomputer* (New York: North Point Press, 2004).

36   Ibid., 259.

37   Spencer Weart, *The Discovery of Global Warming* (Cambridge: Harvard University Press, 2003), 191.

38   Ibid., 191, 198.

39   Brody, *The Other Side of Eden*, 43–44.

40   Ibid., 96.

41   Earthtrends, "CO_2: Cumulative Emissions 1990–2000,"

World Resources Insitutute, http://earthtrends.wri.org (accessed 23 November 2005); United Nations Framework Convention on Climate Change, *Key GHG Data*, 2005, http://unfccc.int.2860.php.

SILA WISDOM FOR A TIME OF CHANGE

Under the leadership of Conservative Prime Minister Stephen Harper, the Canadian government has expressed significant interest in the changes that are impacting the northern ecology of the polar bear and Inuit culture. In his first year of power, Harper held a news conference in Churchill, Manitoba, also on the west Hudson Bay coast, to announce that the melting of the Northwest Passage has highlighted the importance of enhancing "our knowledge of, and presence in, the region."[1] Here and in later announcements Prime Minister Harper stated that because "use it or lose it is the first principle of sovereignty," his government would defend Canada's claim by increasing "scientific inquiry and development" in the North. What is interesting about his concern is that the issue of climate change was never mentioned once, despite the rising interest in a Northwest Passage being related to its odd melting. This omission is consistent with the press release that came out of Minister of Environment Jim Prentice's office following the Polar Bear Roundtable. Though it stated that the wealth of knowledge exchange will inform "decisions related to the conservation and management of the polar bear," it did so in a way that neglected to once mention the word "climate."[2] In contrast to the Conservative government approach that will be further clarified in later chapters, many Inuit have been connecting their rising polar bear *irksi* and the melting Northwest Passage with the warming of northern weather, or *Sila*.

Much of my early e-mail correspondence with Jaypeetee Arnakak centred on the relevance of *Sila* to northern warming. Trying to give me a broad sense of this term, he described *Sila* as an ever-moving and immanent force that surrounds and permeates Inuit life, with it most often being experienced in

the weather.[3] Contrasting his *Inuit Qaujimatuqangit* (IQ) view
of *Sila*, I brought knowledge to our dialogues that was derived
from two largely divided academic disciplines. At one end were
Inuit ethnographies that began in the opening decades of the
twentieth century with researchers like Knud Rasmussen meeting Inuit such as Najagneq. While journeying with the Danish
Fifth Thule Expedition across the Canadian Arctic in search of
Inuit worldviews, he heard of *Sila* described as "a strong spirit,
the upholder of the universe, of the weather, in fact all life on
earth."[4] Ethnographic sources like his have depicted *Sila* as the
spirit of the air,[5] a mystic power permeating all of existence and
a god-like "Supreme Being."[6] Also informing my knowledge
was the more recent climate research on IQ and northern warming that consistently refers to *Sila* as a direct translation for
weather.[7] For example, the research by Natasha Thorpe found
Kitikmeot elders to use *Hila*—a regional variation on *Sila*[8]—as
a translation for weather.[9] These diverging disciplinary interpretations of *Sila* reflect a significant issue that can block the
kind of intercultural and interdisciplinary dialogue I propose
in this book. In the words of Dyanna Riedlinger and Fikret
Berkes, the central problem in conducting such research is that
there is a lack of "conceptual frameworks on how to bridge the
gap between Inuit knowledge and western science."[10]

By the time I boarded a plane to Chesterfield Inlet, Nunavut,
Arnakak's correspondence on *Sila* had clarified the importance
of attempting to traverse these interdisciplinary and intercultural gaps. Upon my arrival, the town's office recommended
Simionie Sammurtok as a workshop co-facilitator and translator because of his concern about the North's changing *Sila* and
his leadership as president of the local Hunter and Trapper
Organization. As we met the afternoon before the workshop
to talk about it and what it could look like, Sammurtok talked
about the increasingly unpredictable *Sila* in relation to hunter
uncertainty, shifting polar bears, and various stories.

When the workshop began the next day, the two of us introduced our concerns about northern climate change and the kind of knowledge-sharing we were aiming for over the next two days. Sitting with us around the table were Andre Tautu, Eli Kimmaliardjuk, Louis Autut, Casemir Kriteedhuk, Elizabeth Tautu, Bernie Putulik, and Mark Amarok—a group of elders, hunters, and a Northern Ranger. They talked of many signs of climate change, including irregular polar bear behaviour, melting ice, and the impact of declining inland water levels on spawning Arctic char. Some spoke about a northern migration of insects, birds, and animals from the treeline. Others referred to the increased frequency of extreme cold days in winter, hot days in summer, and snowstorms. Almost all talked of *Sila's* warming and the resulting change in seasonal timing, which together were affecting birds, reducing the Northwest Passage's sea ice, and increasing polar bear frequency around the community. As I already touched upon in the Introduction, they also spoke of the significant impact these changing signs were having on their culture and IQ. While the next chapter will re-engage Canada's Conservative party's climate politics, I want to begin here with an IQ-Western dialogue focused on *Sila's* northern warming.

BEYOND SILA'S COLONIAL NARROWING

My introduction to *Sila* occurred while reading the ethnography of Knud Rasmussen in an anthropology course during the early 1990s. It was while interviewing the Inuit shaman Najagneq that he heard *Sila*, the spirit of the universe, described in the following way:

> [*Sila* is] so mighty that his speech to man comes not through ordinary words, but through storms, snowfall, rain showers, the sea, through all the forces that man fears, or through sunshine, calm seas or small, innocent children ... When times are good, *Sila* has nothing to say to mankind. He has

disappeared into his infinite nothingness and remains away as long as people do not abuse life but have respect for their daily food. No one has ever seen *Sila*. His place is so mysterious that he is with us and infinitely far away at the same time.[11]

It was almost a decade later, and six months prior to Arnakak's Toronto lecture, that *Sila* surprisingly presented itself again in a film documentary on Inuit observations of climate change called *Sila Alangotok*, or *Weather is Changing*.[12] As with Natasha Thorpe's findings, these researchers state that "observations of climate change by people in Sachs Harbour are based on their knowledge of the weather, or *Sila*."[13] Two years after watching these Inuit from the Northwest Territories talk about various signs of change, I found myself in Chesterfield Inlet hearing many similar observations of a warming and unpredictable *Sila*.

Early in the workshop, Casemir Kriteedhuk spoke of the changing *Sila* in a way that resonated with Tautu's introductory concern about IQ's future:

> They use to know how the *Sila* is changing. They would say something about it and know ahead. But now it is so different, you cannot say what is going to happen in an hour or later in the day. It is not like a hundred years ago when IQ could predict ... Right now it is so different. You will get good *Sila* in the morning and go out on the land because of that, and then get stuck on the land because of the *Sila* changing.

Over the two days of the workshop, I heard many variations on such an assessment, with the general sentiment being that *Sila* changes had become fast and extreme. These voices were as certain something was happening as those in *Sila Alangotok*. Interestingly, these changes were described as not only occurring in the external world, but were also increasing the uncertainty of how IQ was to approach *Sila*.

While Riedlinger and Berkes note that in IQ, "*Sila*, means 'weather', and there is no term to distinguish between weather and climate,"[14] Western knowledge makes a clear distinction between weather and climate through its respective disciplines of meteorology and climatology. The *Oxford Dictionary of Weather* describes "climate" as using statistical information about "the variability of weather conditions prevailing in a particular region or latitude zone over a specific period of time," and defines "weather" as the regional impact of temperature, humidity, wind, and pressure.[15] Reflecting upon the interconnections between climate and weather, Michael Collier and Robert Webb describe the earth as "a weather machine" that redistributes the sun's energy through atmosphere, land, and ocean in the immediate and localized weather and broader climate.[16] The climate's tapestry is not only temporally longer than the weather, but it is also spatially broader, such that a region's latitude combined with geological factors, such as being inland or coastal, can impact local changes in weather.[17] With the climate's global warming, northern *Sila* has been undergoing dramatic changes because of the region's proximity to shifting ice and oceans. But as the Rasmussen quote alludes to, the uncertainty of *Sila*'s changes may go far beyond the conceptual power of scientific climatologists and meteorologists.

Thinking about the 1999 release of the Tlingit ancestor *Kwä-day Dän Ts'ínchi*, or Long Ago Man Found, from a melting alpine glacier in the northwest, Julie Cruikshank writes that this 550-year-old ancestor has reinvigorated oral histories that tell of his "appearance, his contribution to science, his ceremonial cremation" and the required etiquette for maintaining "balance in a moral world."[18] As the northern warming has released *Kwäday Dän Ts'ínchi* and changed the *Sila* Inuit are use to, many researchers have come to realize that, in Cruikshank's words, "orally transmitted genealogies, scientific research, and archival accounts support one another."[19] Despite this intercultural potential, many researchers like Cruikshank and Wenzel have expressed concerns about the cultural gap in different kinds of knowledge—concerns that raise the possibility of a broader reference for *Sila* than simply the weather. The research of Gita Laidler on northern sea ice changes similarly argues that Western assumptions are limiting collaborative research with IQ. In response, she proposes "that intersecting these distinct, yet complementary, perspectives on sea ice will allow for more comprehensive assessments of community vulnerability and viable suggestions for adaptation to climatic change in high latitudes."[20] These critiques resonate with Shari Fox's analysis that "although many northern scholars acknowledge how deeply Inuit are tied to the land in terms of subsistence, knowledge, culture, and spirituality, they often fail to address the latter two themes in discussions related to environmental change."[21] Though the disciplinary assumptions of Western ethnographers or climate researchers may narrow their respective understandings of *Sila*, these learned predispositions are by no means the only issue.

Offering insight on some related issues is Michele Ernsting's conversation with elder Naqi Ehko, who explained that Inuit depended on *Sila* as a force related to the weather and wisdom.[22] The problem is this "meaning is no longer understood by most Inuit youngsters," for many have lost their ancestral land skills and struggle with Inuktitut.[23] The Inuit philosopher Rachel Qitsualik writes that the original inspiration for *Sila* arose from the conclusion that "life itself was in fact the breath, the *Sila*, and that when the *Sila* was drawn into a body, it was alive and animate."[24] Continuing Ehko's critical train of thought, she argues this understanding of *Sila* is not simply being narrowed by reductionary Western research and the lack of translatable terms.[25] As she states:

Although translated today as "air" or "weather" or even "outside," the modern translations of *Sila* only convey to non-Inuit ideas associated with English words. When I speak of the "air" to a southerner, what immediately comes to his or her mind is the idea of invisible, breathable gas: the nitrogen, oxygen, and other gases that make up Earth's atmosphere. Today, the vast majority of Inuit will think of the same thing, as well as wind and weather.[26]

Her point is that the narrowing of IQ to traditional ecological knowledge, and *Sila* to weather, partakes in a lengthy colonial dynamic that has been progressively narrowing the cross-generational understandings of Inuit and, by association, the scope of Western research conducted with Inuit.

It is significant that the first signs of northern warming documented by Western research as occurring in the latter decades of the nineteenth century almost coincide with the colonial onset of Inuit-Canadian relations.[27] Beginning on June 23, 1870, with the British Parliament's passing of the Rupert's Land Act, the three-year-old nation of Canada was given jurisdiction over all the land from which rivers flowed into the Hudson Bay—a large area that included parts of western Quebec, much of northwestern Ontario, the Plains provinces, and the Inuit Territory today known as Nunavut. The twentieth-century colonial evolution of these intercultural relations had a deep cultural impact that is highlighted in the 1996 Canadian-government-ordered Royal Commission on Aboriginal Peoples. Here it is stated that the colonial "premise of resocialization" was to "kill the Indian in the child" by "severing the artery of culture that ran between generations."[28] The report clarifies that these institutions not only disconnected Inuit from their cultural heritage but stranded them "between cultures, deviants from the norms of both."[29] It is because of the twentieth-century experience of these colonial dynamics that Inuit have experienced a widening generational gap in the IQ understanding of *Sila*, as documented by Qitsualik, Ehko, and others. Interestingly, this cultural change has been followed over the past few decades by a warming *Sila*, thus adding more uncertainty to this northern knowledge.

These climatic and colonial dynamics suggest that an appreciation of IQ's broader sense of *Sila* is being limited in Western research by at least three intercultural power dynamics. Firstly, there are Western climate and ethnographic researchers who project limited disciplinary assumptions that cannot blend ecological and cultural knowledge. Secondly, there is the obscuring and degrading power of various colonial forces on the intergenerational sense of IQ that Qitsualik argues is transforming the Inuit view of *Sila* as "a raw life force that lay over the entire Land; that could be felt as air, seen as the sky, and lived as breath."[30] Finally, there seems to be a connection between these earlier colonial dynamics and the current northern warming that is impacting the capacity of those who still have access to IQ to sense these changes. This issue, which is more fully examined in Chapter 4, was highlighted by both Andre Tautu and Casemir Kriteedhuk when they talked about the way in which Inuit lives are being pressured by cultural and environmental forces into a narrowed IQ understanding of *Sila*'s warming. It is these three dynamics that I tried to circumvent by bringing interdisciplinary research into intercultural dialogue with Arnakak, the participants in Chesterfield Inlet, and other Inuit voices.

SILA AND SILATUNIQ

With a few hours left in the Chesterfield Inlet workshop, I offered Rasmussen's translated words of Najagneq on *Sila* for discussion and listened intently as Andre Tautu spoke: "I don't know. I was raised by my grandparents and I never heard a story about *Sila*. It is not from the Inuit history or stories." He added, "This is rubbish," for there was nothing in his experience that equated *Sila* with an Inuit god or spirit as referred to in Rasmussen's quote. When Arnakak reviewed the workshop transcripts, he felt Tautu was bypassing the philosophical context of *Sila*, which would limit an intercultural appreciation of what was being said. He agreed *Sila* is not an anthropomorphic god that can be gendered as a "he," as it is done in Rasmussen's work, and that *Sila* is also not documented in any Inuit stories or art. But, in his view, *Sila* does have a spiritual reference that is meant to contextualize the physicality of human relations within broader ecological processes like the weather.

This interpretive disjunction highlighted the fact that there is not one IQ worldview, a point I will continue to raise in subsequent chapters. Rather, Inuit interpretations and understandings, especially in a post-colonial context, have the potential to be as divergent as the West's ethnographic and climate research views on *Sila*, and yet they can still be part of the same knowledge tradition. Looking over the workshop transcripts, Arnakak came across another passage that seemed to be highlighting his concern. Simionie Sammurtok was attempting to clarify the meaning of *Sila* by stating, "*Sila*, that is the weather, weather wise, climate." Arnakak felt this was an important statement and explained why:

> I think he's referring to *Sila* as "wisdom" because he's listing what the word means. Climate has to do with cycles of change within *Sila*. *Sila* ... moves and is without form of its own. *Sila* is really a no thing because snow is not *Sila*, rain is not *Sila*, wind is not *Sila*, clouds are not *Sila*—they occur in *Sila* ... *Sila* is like empty space.[31]

Contrary to simply forgetting Rasmussen and other ethnographies, Arnakak's correspondence brought their understandings into dialogue with contemporary IQ to clarify a post-colonial sense of *Sila* within a warming North. It is an approach that can broaden the intercultural and interdisciplinary dimensions of climate research.

Following one of our exchanges, I sent Arnakak an article by Daniel Merkur that reviews ethnographic conceptions of *Sila* over the twentieth century.[32] Classifying *Sila* as the indweller of the air, Merkur explains that these ethnographies suggest Inuit traditionally viewed indwellers as "living presences" that "manifest themselves in the miraculous events of the physical world, the unbidden inspiration of the mind, the dreams of sleep, and the visions of ecstasy."[33] Neither supreme beings nor "personifications of natural forces," indwellers "are the powers that constitute nature" as unpredictable beings "that may be benevolent, neutral, or actively malevolent as the mood strikes it."[34] His research also found that as the air indweller, *Sila* has a role in punishing taboo violations through the use of storms and other weather anomalies. Finally, as if responding to Andre Tautu's comment concerning the lack of *Sila* stories, Merkur proposes this may be the case "because many are comic tales and/or negative examples," and as such, "are not told about *Sila*" directly.[35] Responding to the Merkur article, Arnakak wrote that he liked most of the reviewed subject matter but felt that Merkur ultimately "emphasizes the wrong things or in an inappropriate way."[36]

The first problem he saw was the continued use of a gendered "he" when referring to *Sila*, as is common throughout ethnographies like that of Rasmussen and later interpreters such as Merkur. It is more appropriate in Arnakak's view to refer to *Sila* as an "It" that is capitalized to emphasize a nonobjectified reference. In other words, *Sila* as It reflects a sentient and animate force that can potentially respond to other powers and beings. The indigenous scholar George Tinker similarly asserts that indigenous terms are not easily translated into English without losing the breadth of their specific cultural meaning.[37] He views God as one particularly problematic term, for Christian conceptions often carry the baggage of humanlike attributes, such as the emotions of love and anger, as well as being gendered. While Native American traditions offer a diversity of cultural views on God, Tinker explains what they have in common is a characterization that lacks gender and purely human motives. In his words, "what Christians would refer to as God is understood as a spiritual force that permeates the whole of the world and is manifest in countless ways in the world around us at any given moment and especially in any given place."[38] Arnakak said something similar when reminiscing to me about his father and *Sila*: "He only knew that God (*Sila*) is a living, breathing immanence—a reality, not myth—and his worship was in the active partaking of that being."[39] These indigenous views suggest the West's historic grounding in an anthropomorphizing faith has led many ethnographers and religious scholars to give *Sila* a gendered reference that truncates Its animate and more ecologically immanent qualities.

Moving beyond the gendered representation of *Sila*, Arnakak also had problems with some of Merkur's interpretations and their implications. For example, Merkur interprets the phrase "*Sila*up inua" as meaning the "owner of *Sila*," that someone has taken hold of *Sila*. In contrast, Arnakak corrects Merkur's anthropocentric interpretation of this phrase, stating what it actually means is "the person is owned by *Sila*."[40] He explains that *Sila* is not our personification, as represented in much of the ethnographic literature, but "we are the personification of *Sila*." One of his articles focuses on this exact issue by first outlining the importance of recognizing *Sila* as a common root word in Inuktitut that, when combined with other terms, can take on various references, including the delineation between the being of *Sila* as "*Silarjuaq*" and a wisdom that can attend to "It" as "*Silatuniq*." To understand what Arnakak is getting at here, it is helpful to first consider the way in which IQ linguistically joins *Sila* together with various root terms or morphemes to offer different meanings.

In the *Inuktitut Living Dictionary* there are 217 cited cases where *Sila* is joined with other root terms to provide various meanings, many of which concern the weather.[41] For example, *Silaaqsiaq* translates as "darken or change in outside air," while *Silarqiqpuq* means "calm without wind." Adam Qavviaktoq used a regional variation of *Sila* in combination with other terms when talking with an ethnographer about consulting shamans in times of bad weather by using the term *hilaqirhain-ahuaq*, or "he must make the good weather come."[42] Returning to Arnakak's reference to *Silatuniq* as wisdom, the Inuktitut meaning for *tuniq* is "the first peoples who were in this northern land before the Inuit." When combined with *Sila*, it comes to refer to a practical ancestral wisdom for living within this land. Meanwhile, the root term *rjuaq* refers to something big, great, or large, and when combined with *Sila*'s reference to the substance of life, it results in the following description by Arnakak:

> *Silarjuaq* is without a creator. Beings—whether they be animal, human or spirit—become and pass away within it: *Silarjuaq* just is. *Silarjuaq* is also in a state of constant flux and change—reflecting the human mind ... To outsiders (e.g. Rasmussen), *Silarjuaq* would be said to be anthropomorphized, but some Inuit would say that it is we living beings that are *Silarjuapomorphized*. *Silarjuaq* has natural rhythms and cycles as seen in the changing seasons ... everything is mutable—only sentience, order and change are constant.[43]

His further wedding of *Sila* with English terms in the idea of *Silajuapomorphization* reflects an IQ appreciation of the way in which the broader living context informs human life, rather than reality being secondary to one's cultural conceptions.

This interconnecting of "*Sila* without" and "*Sila* within," of *Silarjuaq* and *Silatuniq*, is not unique to IQ. Describing the Navajo view of air, eco-philosopher David Abram states that they insist "the 'Winds within us' are thoroughly continuous with the Wind at large," thus suggesting the "mind *is not ours*, is not a human possession."[44] Closer to northern Inuit lands, Richard Nelson describes the *Koyukon* as personifying the weather, with an awareness that responds to people, and can be manipulated by those who understand its essence.[45] Offering a sense of this intercultural connection, Tinker explains that many kinds of indigenous knowledge experience divinity "as a spiritual force that permeates the whole of the world."[46] This seems consistent with Arnakak's view that *Silatuniq* is needed to understand *Sila* and, in the current context, Its northern warming because, in IQ, the mind can affect the world and the world can affect the mind. Arnakak adds that this approach to *Sila* as *Silarjuaq*, a sentient spiritual consciousness, challenges basic Western assumptions "in a systemic way because it forces one to think of responsibility to self and the other," while also contextualizing "the self into its environment."[47]

In much the same spirit as Qitsualik's reference to a colonial narrowing of *Sila*, Abram describes a history of Western knowledge becoming increasingly anthropocentric and disciplined in its understanding of the human relation to the air, weather, and climate. He writes that in the West's past, the air was understood to be "the invisible wellspring of the present," and was experienced as mediator "between *seen* and the *unseen*."[48] Offering a similar analysis, religion scholar Robert Torrance states the ancient Latin *spiritus* associated the breath with an animating force that spiritually united inner and outer.[49] According to Abram, such an inspired sensibility was more real for oral peoples than the "abstract dualism between sensuous reality as a whole and some other, utterly non-sensuous heaven" that has evolved in monotheistic religions like Christianity.[50] This dissociation of the air's knowable and unknowable qualities continues today in its scientific conception as an empty space that denies "our thorough interdependence with the other animals, the plants, and the living land that sustains us."[51] In a comparable IQ view, Arnakak not only describes "a mature mind" as one that can reflect on its contextual immersion, but also argues that because reductionary science has forgotten this reality, it "is lost in its angst affecting everything everywhere."[52] The way out of this narrowing Western dynamic is, for Abram, dependent upon once again fostering an awareness of the surrounding invisible air as "a potentized field of intelligence in which our actions participate" rather than being a passive backdrop for human actions.[53] This is not a romantic return to oral lifeways and knowledge but a wise attempt to unite "our capacity for cool reason with those more sensorial and mimetic ways of knowing, letting the vision of a common world root itself in our direct, participatory engagement with the local and the particular."[54]

Though Torrance's cross-cultural analysis of *Sila*-like sensibilities supports the phenomenological view of Abram, he contrasts it by suggesting there is something in the West's scientific search for objective knowledge that partakes in this "universal activity by which humanity is in large part defined as human."[55] Despite its aim for objectivity, there is at the core of science a self-transcending approach that recognizes reality's indeterminate nature and the limits to complete knowledge. This indeterminism can also be seen to underlie the uncertainties of climate research, for, as David Demeritt writes, "neither the idea of a 'global climate' nor the phenomena that it designates are conceivable apart from the world-shaping networks of social practices, standardized instruments, orbiting weather and communication satellites, and computer models through which they are made manifest."[56]

Further highlighting our relation to northern warming, climate scientist George Philander describes "the music we call weather" as a symphony of forces that includes the sun, planetary orbits, oceans, atmosphere, air masses, ice, regional topography, plants, animals, and humanity's greenhouse gases.[57] When brought together with Abram's analysis, this weather music takes on more spiritual dimensions as it becomes intimately connected to the cycling inhalation of animals and exhalation of plants. As Abram states, the air as the invisible source for all of life's nourishment is the "most intimate absence from whence the present presences, and thus a key to the forgotten presence of the earth."[58] The problem is that today's climate research agenda largely does not deem it highly valuable or adaptive to think about such ethical and spiritual dimensions of *Sila's* changing presence. To begin assessing the potential of more fully incorporating such spirited intercultural understandings in interdisciplinary climate research, we need to delve deeper into ethnographies like that of Rasmussen on Inuit shamanism and the *Silatuniq* required for northern living.

SHAMANIC INITIATION

The decades following Rasmussen's exchange with Najagneq saw many religion and mythology scholars, like Joseph Campbell and Mircea Eliade, consider the cross-cultural relevance of shamanism. In the popularized and sometimes contentious work of Campbell, the shaman's *Silatuniq* was described as arising out of an initiation experience into a spiritual reality that is interpreted by everyday consciousness as the "secret cause" of tragedy and worldly fear.[59] To understand his perspective, we need to follow Campbell's analysis into an earlier part of the exchange when Rasmussen had just realized Najagneq often fabricated stories and used trickery in rituals to mystify community members. Because of this awareness, Rasmussen asked the shaman if there was anything he believed. Najagneq responded: "Yes, a power that we call *Sila*, one that cannot be explained in so many words. A strong spirit, the upholder of the universe," and, as quoted earlier, he goes on to recite the power of *Sila* as a mystery that "is with us and infinitely far away at the same time."[60] These words resonated with those of other shamans Rasmussen met. For example, Igjugarjuk from Baker Lake, just inland from Chesterfield Inlet in Nunavut, told him that the "only true wisdom lives far from mankind, out in the great loneliness, and it can be reached only through suffering."[61] This human and cultural need to cosmologically connect knowledge with a power that transcends the sufferings and joys of phenomenal life is for Campbell a serious methodological lacuna for today's overly rationalized Western culture. If his perspective is valid, it would suggest there is also a spiritual block involved in Western misunderstandings of IQ concepts like *Sila*.

In Campbell's research, it is the process of coming into tune with such a reality as *Sila* that requires the shaman to undergo, for community and culture, the suffering of a solitary initiatory experience. While Torrance describes such a spiritual quest as an ever-recurring process of trying to understand "a future goal that can neither be fully known nor finally attained,"[62] Mircea Eliade similarly writes that the universal search for spiritual insight "always entails death to the profane condition, followed by a new birth."[63] The suffering journey for Igjugarjuk, as told to Rasmussen, began with an initiation that entailed being left alone on the tundra in a small hut just big enough to sit cross-legged. Perched on a small hide upon the snow, he was forbidden to touch anything, had limited food and water, and was far from his community. Igjugarjuk described those thirty days of cold and fasting as being so severe that he "sometimes died a little." During his initiatory death, Igjugarjuk thought only of the Great Spirit, endeavouring to keep his memory free "of human beings and everyday things."[64] Through such initiations, shamans like Igjugarjuk contacted spirit helpers and found in the "great loneliness" a *Silatuniq* that, in Arnakak's words, does not seek "to remedy but to help rebalance, to massage the individual to its natural equilibrium with the environment."[65] In the *irskina* experience of being spiritually devoured by *Sila*'s "great loneliness," the shaman could manifest for Inuit a contextualizing *Silatuniq*.

The Western assumptions that block access to such a wise

initiation are highlighted in the romantic approach to northern realities like the iconic Northwest Passage and the polar bear. In his popular book, *Arctic Dreams*, Barry Lopez finds that as the Western view on the Arctic changed from one of an inhospitable place to a region of limited economic value, the vision of the polar bear also transformed from one of wild indifference to "a vaguely noble creature."[66] He writes that in the modern Western mind, the polar bear wanders "a desolate landscape, saddled with melancholy thoughts" much like an estranged and self-absorbed romantic.[67] It is ironic that both George Wenzel and Canadian social critic Renée Hulan have critiqued Lopez for being exactly this kind of a romantic who could not truly engage the northern reality of the Inuit. The extent to which the writing is confined to stories of southern scientists and professionals leads Wenzel to conclude that Lopez probably "found Inuit too ordinary" for his thoughts on the North.[68] Similarly, Hulan has more broadly critiqued non-Inuit writing on the North, like that of Lopez, as being preoccupied with two interrelated romantic assumptions that can be seen as both informing the Conservative government's iconic approach on polar bears and blocking Western climate research's appreciation of the need for *Silatuniq*.

While the first romantic assumption holds that Inuit have access to the Stone Age past as a hunting society, Hulan explains the second assumes that Inuit can offer Westerners access to an exotic northern spirituality.[69] She writes that for many Canadian and, more broadly, Western writers, "Inuit have been imagined as ideal Canadians, as those who can pass on the 'autochthonous claim' to both the land and the north."[70] In contrast, her research on Inuit self-representation reveals a contextualized definition of the individual as "identified with the rest of the community," rather than embodying a rugged individualism that is essential to Western thought.[71] Highlighting the dual romantic nature of this contrast, Wenzel writes that Westerners undertake Arctic expeditions by making elaborate military preparations in a homeland that "Inuit survive through arcane customs and skills that fascinate and repel us all in one."[72] He adds the important point that "Inuit serve as an allusion to a world inconceivably more difficult than the privileged one we know." In looking at Inuit self-representation, it becomes apparent to Hulan that Western romantics imagine fleeing north into "a perfect solitude, a perfect separation from social bonds, a rupture with the social contract, which, just like an ideal masculine identity, can never really be."[73] This is, in her view, the romantic consciousness of Canadians who like to see themselves as "similarly rugged and individualistic or, negatively, lacking and needing an understanding of its true nature as rugged and individualistic."[74] The apparent individualism of an Inuit shaman entering the "great loneliness" is not the same as the self-absorbed and individualistic romantic. As Eliade's research clarifies, the shamanic *Silatuniq* that would arise from their *irksina* initiation would ideally help Inuit communally adapt to problems related to hunting, sickness, and bad *Sila*.

Whether the analyses of Campbell and Eliade on shamanism and *Sila* were similarly flawed by romantic cultural beliefs was a question I posed Arnakak during one exchange. In his opinion, the representations were incisive:

> I truly believe in this *Sila* and the means with which Inuit shamanism accessed its depths and breadth through suffering and fasting. It is through suffering that the phenomenal self lets go and equanimity is achieved, clarity is achieved. Nature is indifferent; it cares nothing for our limited conceptions of 'good' and 'bad,' 'evil' and 'beneficence.' This insight can either kill us or liberate within us unbound creativity.[75]

Arnakak suggests the suffering of shamanic initiation is a process for letting go of the cultural beliefs that limit participation in social realities that transcend and, eventually, consume our bodies. Continuing this train of thought, he writes, "Shamans are the sensitive people in so far as they're able to sense and perceive what normal people cannot or will not face."[76] Though the shaman's *Silatuniq* allows the community to be brought into a *Sila* relationship that "is fundamentally an ethical one, not economic," Arnakak clarifies that the difficulty is that this *Silatuniq* may contradict dominant cultural assumptions for everyday living.[77] Beyond the narrowing dynamics of cultural assumptions, colonial pressures, and expanding climate impacts, this dialogue suggests a fourth dynamic is limiting the breadth of interdisciplinary and intercultural research on *Sila's* northern warming: the West's rational rejection of shamanic or spiritual wisdom for socially contextualizing knowledge.

CONCLUSION: TOWARD WISE CLIMATE RESEARCH

Harkening back to the Introduction, Eliade tells us that some shamanic initiations were mediated by a polar bear that tears, devours, and then "becomes the future shaman's helping spirit."[78] One Labrador Inuit actually described the Great Spirit, who Najagneq connected with *Sila*, as appearing in the form of an *irksina* polar bear that "devours the aspirant."[79] The relevance of such a terrifying initiatory experience to Western environmental research was highlighted by eco-feminist Val Plumwood in the early 1990s, after she endured the actual physical experience of "being prey" to a large predator.[80] Though she survived the incident, it is clear Plumwood did not come out of the experience untransformed. Looking back on the machinations of her thoughts as the event unfolded, she writes that "the mind can instantaneously fabricate terminal doubt of extravagant, Cartesian proportions: this is not really happening, this is a nightmare, from which I will soon awake."[81] She would later connect the lack of conservation success over the twentieth century to just such a "rationalist failure to situate the human in ecologically embodied and socially embedded ways."[82] This rational delusion was for Plumwood personally shattered when the inevitable reality of death was forced upon her consciousness. It was then that she saw, for a moment, the world from the "outside" as a bleak reality that was indifferent to her needs, passions, and beliefs. The long human history of being vulnerable prey, as well as powerful predator, has in her view allowed us to evolve a capacity to scent danger that is still known "to certain indigenous cultures, but lost to the technological one which now colonises the earth."[83] In this uncertain context, she argues that intuitive sensibilities—like that of the Inuit polar bear *irskina*, colonial *ilira*, and shamanic *Silatuniq*—are valid in that they teach us about the power of natural processes to resist limiting cultural assumptions. While she hopes "that it does not take a similar near-death experience to instruct our culture in this wisdom," the challenge of today's global climate changes to both Inuit and Western knowledge systems can be seen as just such a terrifying initiation into a new, yet ancient, way of recognizing our worldly context.

The initiation of Western climate research into such a wise approach conceptually begins, for me, over a half-century ago with the elucidation of what can be seen as the earth's climatically changing respiration. This scientific view on the planet was unveiled by Charles Keeling in the 1950s after years of research on carbon dioxide (CO_2) concentrations in the high altitude of Mt. Mauna Loa in Hawaii. Documenting trends that came to be known as the Keeling Curve, his graph revealed a decrease in CO_2 concentrations each spring as the sprouting greenery of temperate regions extracted CO_2 from the atmosphere, followed by an autumn rise as decomposing greenery exhaled the CO_2.[84] What was most curious to Keeling and subsequent climate researchers was the general global trend of an increasing exhalation of CO_2 over the span of the twentieth century.[85] Almost in lockstep with both the West's rising fossil fuel combustion and the Inuit colonial experience has been the levels of CO_2 that rose from about 295 parts per million (ppm) in 1900 to 315 ppm in

1950 and about 360 ppm in 1995.[86] In 2004 Mauna Loa revealed an increase of 3 ppm from the previous year, bringing its concentration to 379 ppm, with the Intergovernmental Panel on Climate Changes's (IPCC's) 2007 report stating that this level "exceeds by far the natural range over the last 650,000 years (180 to 300 ppm)."[87] Coinciding with this heightened planetary breath has been a global warming trend that has seen eleven of the twelve years preceding 2005 ranked among the top ten temperatures recorded since 1850.[88] It is increasingly clear that the greenhouse gas emissions of industrial activities have been central to this heightening of the planet's seasonal respiration and, consequently, *Sila's* northern warming. The West is in a sense being initiated into the need for a *Silatuniq* that can, in Arnakak's words, inquire into "the context and consequence of applying knowledge and/or how our interacting with the surround affects that surround."[89]

In such an uncertain context, Arnakak states IQ looks upon the sharing of experiential knowledge as the "most reliable and relevant" approach for determining how people can act.[90] Interestingly, the IPCC invokes a process of interdisciplinary knowledge-sharing, though it is largely limited to rational models and research observations that are predicated upon Western scientific and economic assumptions.[91] In fact, Weart argues that despite the evolving changes in climate research from the time of Keeling to the international efforts of the IPCC, it is still too limited by disciplinary specialists during a time of "labor to understand increasingly complex topics."[92] He adds that since the "Earth's climate system is so irreducibly complicated," there is a need for research to become equally intricate in its intertwining of the physical and social sciences.[93] This is similar to the post-colonial critiques of Cruikshank and Wenzel that suggest the ethical and spiritual dimensions of IQ and *Sila* require researchers to broaden their interdisciplinary approach. If it is the case that specific shamanic practices are required to reveal *Silatuniq*, then it may be that an interdisciplinary research en-

deavour that does not value spiritual understandings and practices will have difficulty contemplating what Arnakak referred
to as the *Silarjuapomorphization* of a changing climate. *Silatuniq* may be just as important as today's more common research
preoccupation with IQ's traditional ecological knowledge, for
wisdom may be the most fundamental factor that interdisciplinary climate research needs to clarify as it responds to *Sila's*
warming, shifting polar bear behaviour, and a melting Northwest Passage. The next chapter will look at the potential of such
a *Silatuniq*-inspired climate research endeavour in the context
of Canadian political economic and cultural dynamics.

ENDNOTES

1   Stephen Harper, "Prime Minister Harper Bolsters Arctic Sovereignty with Science and Infrastructure Announcements," Media Release, Churchill, MB, 5 October 2007, http://pm.gc.ca/eng/media.asp?id=1843.

2   Environment Canada, "Minister Prentice Highlights Progress Made at Polar Bear Roundtable," Media Release, 16 January 2009, http://www.ec.gc.ca/default.asp?lang=En&n=714D9AAE-1&news=24AABBD9-00C3-4E80-9517-2D37013C5FAF.

3   E-mail correspondence, 24 January 2005.

4   Knud Rasmussen and Hother Berthel Simon Ostermann, *The Alaskan Eskimos as Described in the Posthumous Notes of Knud Rasmussen* (Report of the Fifth Thule Expedition, 1921–1924, Copenhagen: Gyldendal, 1952), 97–98.

5   For example, Knud Rasmussen, *Across Arctic America: Narrative of the Fifth Thule Expedition* (New York: Greenwood Press, 1969).

6   For example, Kaj Birket-Smith, *Ethnography of the Egedesminde District: With Aspects of the General Culture of West Greenland* (New York: AMS Press, 1976); Wilhelm Schmidt, *The Origin and Growth of Religion: Facts and Theories* (New York: Cooper Square Publishers, Inc., 1972); Raffaele Pettazzoni, *The All-knowing God* (London: Methuen and Co., Ltd., 1956); Åke Hultkrantz, "Les Religions du Grand Nord Américain," in I. Paulson, Å Hultkrantz, and K. Jettmar, eds., *Les Religions Arctiques et Finnoises* (Paris: Payot, 1965).

7   For example, Dyanna Jolly et al., "We Can't Predict the Weather Like We Used To," in I. Krupnik and D. Jolly, eds., *The Earth Is Faster Now: Indigenous Observations of Arctic Environmental Change* (Fairbanks, AK: Arcus, 2002); Natasha Thorpe et al., "Nowadays It Is Not the Same," in I. Krupnik and D. Jolly, eds., *The Earth Is Faster Now: Indigenous Observations of Arctic Environmental Change* (Fairbanks, AK: Arcus, 2002);

D. Riedlinger and F. Berkes, "Contributions of Traditional Knowledge to Understanding Climate Change in the Canadian Arctic," *Polar Record* 37, no. 203 (2001): 315–328.

8   For a discussion on regional variations of *Sila* see Daniel Merkur, "Breath-soul and the Wind Owner: The Many and the One in Inuit Religion," *American Indian Quarterly* 7, no. 3 (1983): 23–39.

9   Thorpe et al., "Nowadays It Is Not the Same."

10   Riedlinger and Berkes, "Contributions of Traditional Knowledge," 316.

11   Rasmussen and Ostermann, *The Alaskan Eskimos.*

12   *Sila Alangotok*, directed by Bonnie Dickie et al., International Institute for Sustainable Development, Winnipeg, MB, 2000.

13   Jolly et al., "We Can't Predict the Weather Like We Used to," 95.

14   Riedlinger and Berkes, "Contributions of Traditional Knowledge," 317.

15   S. Dunlop, *A Dictionary of Weather* (Oxford: Oxford University Press, 2001), 47, 250.

16   Michael Collier and Robert H. Webb, *Floods, Droughts, and Climate Change* (Tucson: University of Arizona Press, 2002), 32; also see S. George Philander, *Is the Temperature Rising? The Uncertain Science of Global Warming* (Princeton: Princeton University Press, 1998), 104.

17   Dunlop, *A Dictionary of Weather.*

18   Julie Cruikshank, *Do Glaciers Listen? Local Knowledge, Colonial Encounters, and Social Imagination* (Vancouver: UBC Press, 2005), 250.

19   Cruikshank, *Do Glaciers Listen?* 40.

20   Gita J. Laidler, "Inuit and Scientific Perspectives on the Relationship between Sea Ica and Climate Change: The Ideal Complement," *Climatic Change* 78 (2006): 407–444.

21   S. Fox, "These Are Things that Are Really Happening: Inuit Perspectives on the Evidence and Impacts of Climate Change

in Nunavut," in I. Krupnik and D. Jolly, eds., *The Earth Is Faster Now: Indigenous Observations of Arctic Environmental Change* (Fairbanks, AK: Arcus, 2002), 45.

22   Michele Ernsting, "The Meaning of Sila," *Radio Netherlands*, 21 December 2001.

23   Ibid.

24   Rachel Attituq Qitsualik, "Sila," *Nunatsiaq News*, 7 July 2000, http://www.nunatsiaqonline.ca; Rachel Attituq Qitsualik, "Word and Will, Part Two: Words and the Substance of Life," *Nunavut Edition*, 12 November 1998.

25   Ibid.

26   Ibid.

27   Martin J. Siegert and Julian A. Dowdeswell, "Glaciology," in M. Nuttall and T. V. Callaghan, eds., *The Arctic: Environment, People, Policy* (Amsterdam: Harwood Academic Publishers, 2000).

28   Royal Commission on Aboriginal Peoples, *Report of the Royal Commission on Aboriginal Peoples* (Ottawa: Indian and Northern Affairs Canada, 1996), http://www.ainc-inac.gc.ca/ch/rcap/sg/sgmm_e.html (accessed 10 November 2005).

29   Ibid.

30   Qitsualik, "Sila"; Qitsualik, "Word and Will."

31   E-mail correspondence, 3 December 2004.

32   Merkur, "Breath-soul and the Wind Owner."

33   Daniel Merkur, *Powers Which We Do Not Know: The Gods and Spirits of the Inuit* (Moscow, ID: University of Idaho Press, 1991), 255.

34   Ibid., 29.

35   Merkur, "Breath-soul and the Wind Owner," 35.

36   E-mail correspondence, 25 January 2005.

37   George E. Tinker, "Community and Ecological Justice: A Native American Response," in D. B. Conroy and R. L. Petersen, eds., *Earth at Risk: An Environmental Dialogue Between Religion and Science* (Amherst, MA: Humanity Books, 2000).

38 Ibid., 244.

39 E-mail correspondence, 26 May 2004.

40 E-mail correspondence, 24 January 2005.

41 *Inuktitut Living Dictionary*, http://livingdictionary.com (accessed 12 July 2006).

42 John Bennett and Susan Diana Mary Rowley, *Uqalurait: An Oral History of Nunavut* (Montreal: McGill-Queen's University Press, 2004), xxi.

43 Jaypeetee Arnakak, *A Case for Inuit Qaujimanituqangit as a Philosophical Discourse* (Iqaluit, NU: JPT Consulting, 2004), 2.

44 David Abram, *The Spell of the Sensuous: Perception and Language in a More-than-Human World* (New York: Pantheon Books, 1996), 254.

45 Richard K. Nelson, *Make Prayers to the Raven: A Koyukon View of the Northern Forest* (Chicago: The University of Chicago Press, 1983).

46 Tinker, "Community and Ecological Justice," 244.

47 E-mail correspondence, 17 March 2004; 19 March 2004.

48 Abram, *The Spell of the Sensuous*, 254.

49 Robert M. Torrance, *The Spiritual Quest: Transcendence in Myth, Religion, and Science* (Berkeley: University of California Press, 1994), 55–56.

50 Abram, *The Spell of the Sensuous*, 254.

51 Ibid., 260.

52 E-mail correspondence, 2 March 2004.

53 Abram, *The Spell of the Sensuous*, 260.

54 Ibid., 270.

55 Torrance, *The Spiritual Quest*, xii.

56 David Demeritt, "The Construction of Global Warming and the Politics of Science," *Annals of the Association of American Geographers* 91, no. 2 (2001): 312.

57 S. George Philander, *Is the Temperature Rising? The Uncertain Science of Global Warming* (Princeton, NJ: Princeton University Press, 1998).

58 Abram, *The Spell of the Sensuous*, 226.

59   Joseph Campbell, *Primitive Mythology* (Harmondsworth, UK: Penguin Books, 1976).

60   Rasmussen and Ostermann, *The Alaskan Eskimos*, 97–99.

61   Rasmussen, *Across Arctic America*, 83–84.

62   Torrance, *The Spiritual Quest*, 52–53.

63   Mircea Eliade, *The Sacred and the Profane: The Nature of Religion* (New York: Harcourt, Brace, 1959); Mircea Eliade, *The Myth of the Eternal Return: Or, Cosmos and History* (Princeton, NJ: Princeton University Press, 1971); Mircea Eliade, *Shamanism: Archaic Techniques of Ecstasy* (Bollingen Series, Princeton, NJ: Princeton University Press, 1972).

64   Rasmussen, *Across Arctic America*, 83–84.

65   E-mail correspondence, 8 July 2004.

66   Barry Lopez, *Arctic Dreams: Imagination and Desire in a Northern Landscape* (Toronto: Bantam Books, 1989), 113.

67   Ibid.

68   George Wenzel, *Animal Rights, Human Rights: Ecology, Economy and Ideology in the Canadian Arctic* (London: Belhaven Press, 1991), 14.

69   Renée Hulan, *Northern Experience and the Myths of Canadian Culture* (Montréal: McGill-Queen's University Press, 2002), 60.

70   Ibid.

71   Ibid., 86.

72   Wenzel, *Animal Rights, Human Rights*, 13.

73   Hulan, *Northern Experience and the Myths of Canadian Culture*, 151.

74   Ibid., 185.

75   E-mail correspondence, 27 January 2005.

76   Ibid.

77   E-mail correspondence, 12 July 2004.

78   Eliade, *Shamanism*, 44–45.

79   Ibid., 58–59.

80   Val Plumwood, "Human Vulnerability and the Experience of Being Prey," *Quadrant* 39, no. 314 (1995): 29–34.

81   Ibid., 30.

82   Val Plumwood, *Environmental Culture: The Ecological Crisis of Reason* (New York: Routledge, 2002), 26.

83   Plumwood, "Human Vulnerability and the Experience of Being Prey," 34.

84   Tim F. Flannery, *The Weather Makers: How We Are Changing the Climate and What It Means for Life on Earth* (Toronto: HarperCollins Canada, 2006), 25.

85   Flannery, *The Weather Makers*; Spencer Weart, *The Discovery of Global Warming* (Cambridge: Harvard University Press, 2003), 191.

86   John Robert McNeill, *Something New under the Sun: An Environmental History of the Twentieth-Century World* (New York: W. W. Norton and Company, 2000), 109.

87   IPCC, *Climate Change 2007: The Physical Science Basis, Summary for Policymakers, Contribution of Working Group I to the Fourth Assessment Report of the IPCC* (Geneva, Switzerland: IPCC Secretariat/World Meteorological Organization, 2007), 2, 4; IPCC, *Climate Change 2007: Climate Change Impacts, Adaptation, and Vulnerability, Summary for Policymakers, Contribution of Working Group II to the Fourth Assessment Report of the IPCC* (Geneva, Switzerland: IPCC Secretariat/World Meteorological Organization, 2007), 2.

88   IPCC, *Climate Change 2007: The Physical Science Basis*, 2, 4.

89   E-mail correspondence, 18 February 2004.

90   Arnakak, *A Case for Inuit Qaujimanituqangit*, 4.

91   Weart, *The Discovery of Global Warming*.

92   Ibid., ix.

93   Ibid.

RESEARCHING GAIA'S UNCERTAIN CLIMATE

In the lead-up to Prime Minister Stephen Harper's first elected minority Conservative government in January 2006, John Bennett of the Sierra Club raised a common concern among environmentalists that Harper's comments on climate change while in opposition revealed a limited view "of the international situation, a lack of understanding of climate science and a selective use of facts."[1] Three years later, a similar critique was raised by environmental scientists and organizations at the Polar Bear Roundtable. Following the meetings, Andrew Derocher of Polar Bears International argued that despite climate warming being "the main threat to polar bears," the government's position was dominated by questions of sustainable harvest.[2] This critique seems validated by the way in which the government failed to publicly mention climate change in relation to its concerns about either the polar bear issue or a melting Northwest Passage. These are odd omissions considering the extent to which Western research, not to mention *Inuit Qaujimatuqangit* (IQ), suggests these particular issues are related to a northern warming trend that has potentially serious regional, national, and global consequences. My primary concern here is to begin highlighting those political economic and cultural assumptions that are limiting Western climate research in a way similar to the already discussed narrowing of IQ and *Sila*.

An introduction to these issues is provided by David Hallman in his critical analysis of the way in which developing countries have had fewer resources than rich nations for funding the research of the Intergovernmental Panel on Climate Change (IPCC).[3] Though many cultures and nations participate in this research, he points out that the IPCC has been dominated by scientists and political economists whose "analytic design has tended to be dominated by Western economic models, with their emphasis on the monetization of everything possible."[4] A similar analysis of Western environmental research has been made by Wolfgang Sachs. Ever since the 1987 United Nations Brundtland Commission, he writes there has been an increasing focus on utilizing ecosystems science in plans to sustainably develop a planet "whose stability rests on the equilibrium of its components, like population, technology, resources and environment."[5] In fact, Sachs proposes that such an intentional use of science to manage and adjust ecological systems is a strategy that amounts "to completing Bacon's vision of dominating nature, albeit with the added pretension of manipulating her revenge."[6] Just as with the narrowing of IQ and other kinds of indigenous knowledge into traditional ecological knowledge, both Sachs and Hallman suggest environmental and climate management regimes have tended to marginalize both Western ethical and indigenous cultural views that could provide an alternative way of understanding environmental issues like *Sila*'s northern warming.

This rationalizing dynamic has been more recently documented by eco-theologian Anne Primavesi in a way that resonates both with IQ's contextualizing approach to *Silatuniq* and Spencer Weart's conclusion that the future climate "actually does depend in part on what we think about it."[7] Looking at the evolving climate research endeavour, she proposes that these uncertainties are calling us to move our sense of the earth toward an increasingly wise recognition "of our place within it."[8] This linking of climate change to our thought is the starting point of Primavesi's analysis as she leads us through various inquiries on both the present impact of prevailing cultural beliefs and practices, and the potential value of religion in a wise interdisciplinary rethinking of our political economic responses. Though she discusses the role the IPCC has played in opening space for this change of thought, the book's title, *Gaia and*

Climate Change, reflects its indebtedness to James Lovelock's theory on a planetary climate system that, in her words, "binds all our lives and our fates together."[9] While today's climate changes are revealing the importance of transforming our understanding of human-planet relations, Primavesi is concerned that this initiation is not occurring because the focus is still on utilizing Gaia's resources—despite this dominant assumption being at the core of today's disturbances. To consider the West's cultural blocks to including this wisdom, I will use this chapter to outline the evolution of climate research in light of the critical views represented by the other two main parties involved at the Polar Bear Roundtable—Canada's Conservative government and Inuit interests. I will then conclude by beginning to define a Western *Silatuniq* that can add ethical and spiritual dimensions to its well-developed interdisciplinary knowledge of *Sila*'s northern warming and, more broadly, Gaia's climate changes.

REVOLUTIONS IN CLIMATE THOUGHT

A half-century prior to Charles Keeling's discovery of the planet's climatically changing breath, Svante Arrhenius had hypothesized that atmospheric carbon dioxide (CO_2) may increase due to the Industrial Revolution's rising fossil fuel use, thus warming the planet.[10] His 1904 research indicated that a doubling of CO_2 could increase the earth's average temperature by 5° to 6° C, with such a change projected to take three millennia.[11] Future generations would, in Arrhenius's words, "enjoy ages with more equable and better climates, especially as regards the colder regions of the Earth, ages when the Earth will bring forth much more abundant crops than at present."[12] These positive projections have been confronted by many factors, not the least of which was the unforeseen increase of coal production from 1,000 to 5,000 million metric tons and oil production from 20 to 3,000 million metric tons between Arrhenius's time and the late-twentieth century.[13] In contrast to his projections, the IPCC indicates that within two hundred years, concentrations will be two to four times pre-industrial levels.[14] Considering the correlation of past CO_2 levels with global temperatures, it should not be surprising that temperatures have increased by 0.3° to 0.6° C between 1890 and the end of the twentieth century.[15] These projections, which have changed dramatically over the past century, highlight the significant shifts Western research of human-climate relations has undergone—changes that must continue if we are to respond to the present challenge.

Though Western climate thought can be conceived as stretching all the way back to the Greek philosophy of Theophrastus, James Fleming shows in his climate history that today's more rational vision was initiated with the seventeenth- and eighteenth-century Enlightenment ideas of Jean Baptiste Abbé Du Bos, Baron de Montesquieu, and David Hume.[16] This is not to say the specific ideas of these thinkers have survived to the present, for they commonly held a deterministic vision of climate as something improvable through "draining the marshes, clearing the forests, and cultivating the soil."[17] Drawing upon Du Bos's claim that the genius of Western rationality was related to a warming trend, Hume proposed the temperature increase was due to the land being "better cultivated, and that the woods are cleared, which formerly threw a shade upon the earth, and kept the rays of the sun from penetrating."[18] Rationality was connected to a deforesting that influenced the air, weather, and people, and thus it was assumed deforesting colonial lands would also beneficially change their climates and cultures.[19] Emblematic of this thought in North America was Benjamin Franklin's 1763 writing that "cleared land absorbs more heat and melts snow quicker," with the implication also being that these changes would improve the thought of local indigenous and colonials.[20] Some of the Enlightenment ideas that informed these thoughts included the view that nature was deterministic, God was debatable, and applied reason could perfect humanity and the world.[21] As Julie Cruikshank describes in her recent book, *Do Glaciers Listen*, it was this enlightened cultural perspective that informed the earliest relationships of Western researchers with the North's climate, ecology, and indigenous people.

In considering the colonial legacy of Enlightenment research on the Tlingit from North America's northwestern coast, Cruikshank analyzed the historic records and indigenous oral stories concerning the 1786 landing of the French explorer Jean François de La Pérouse at Lituya Bay in the Gulf of Alaska. This exploration was not a purely scientific agenda, for as she writes, La Pérouse had instructions from King Louis XVI "to match in the North Pacific the scientific and cartographic achievements that James Cook had accomplished for England in the South Pacific."[22] Upon arrival, his crew installed a scientific observatory and then proceeded "to measure nature's dimensions, tame its uncertainties, and ascertain its physical attributes."[23] Some of his focus was on the St. Elias mountain where glaciers hung that were changing dramatically due to the cooling of the Little Ice Age, and which he imagined to be "manifestations of the Sublime—great yet terrible, wondrous yet fearsome."[24] Beyond the sublime ice and ecology, La Pérouse also fixed his enlightened gaze on a Tlingit culture that he described as "rude and barbarous, their soil is wild and rugged, they inhabit the country only to extirpate every thing that lives."[25] Following Enlightenment thinkers like Du Bos, Montesquieu, and Hume, he assumed that since nature has the power of "determining what culture is destined to become," it should be possible to civilize the Tlingit through an agricultural transformation of the land.[26] This approach would mark an intercultural dynamic by which indigenous cultures and knowledge were marginalized as superstitious, while the West's scientific research and civilization would be progressively seen as the best way for understanding complex ecological and climate processes.

With the rise of experimental science and national observational systems in the nineteenth century, many of these enlightened assumptions came into disrepute as inexact philosophical musings that, despite questioning Christianity, were mired in cultural beliefs. Directly contrasting the belief that Europe's rational ascendance was related to a warming climate effected by agricultural clearances, the research of Noah Webster at the turn of the nineteenth century found "no evidence for a major climatic change" related to human actions.[27] This scientific change of view is also seen in the glacial research that was of such interest to La Pérouse and which the nineteenth-century Swiss geologist Louis Agassiz transformed significantly. Disagreeing with the dominant Christian belief of his day that patterns of rock and debris on the Eurasian and North American continents were dispersed by the Old Testament flood, Agassiz offered a view that accorded more seamlessly with the accumulating data.[28] Comparing the patterns and grooves occurring both close by and far away from glaciers, he concluded there was at one time a large ice sheet that "extended beyond the shorelines of the Mediterranean and of the Atlantic Ocean, and even completely covered North America and Asiatic Russia."[29] Research has since supported his basic premise that the planet undergoes regular periods of glacial recessions and expansions in relation to the planet's orbital cycling of the sun and other factors. This objective approach to glaciation was part of a general trend to scientifically uncover universal climate laws that transcend limiting cultural assumptions.

A scientific research focus on climate change itself largely begins with Joseph Fourier's 1824 reading of a paper, later published in 1827, on the "rational law of atmospheric motion, ocean motion, change of seasons, and so on—a grand geophysical law confirmed in the laboratory and expressed by calculus."[30] His research on the planetary greenhouse effect would later influence Arrhenius to project that the industrial era could influence this climate system. As he states in his 1904 proposition, "increasing the carbon dioxide content of the atmosphere by burning fossil fuels might be *beneficial*, making the Earth's climate warmer and more equable, stimulating plant growth, and providing more food for a larger population."[31] In contrast to present concerns about the potential negative impacts of high CO_2 emissions, Arrhenius envisioned the fossil-fuel hungry Industrial Revolution as manifesting a warming climate that would stem the eventual cyclical return of the planet to a glacial ice age.[32] This hypothesis that CO_2 emissions could beneficially influence the climate was largely ignored until around the time of Keeling's research when increasingly powerful scientific tools and methods began documenting a warming trend that seemed to begin in the early part of the twentieth century.

The evolving climate research that followed Keeling's discoveries has been described by climate historian William F. Ruddiman as one of four major revolutions in the earth sciences that commonly reveal a much more dynamic planet than previously assumed.[33] The first was James Hutton's eighteenth-century dating of the planet to a more ancient time than that of the six-thousand-year-old story of the Christian churches. As with Agassiz's glacial research, this recognition led to much scientific examination that physically confirmed the earth is several billions years old.[34] The second earth science revolution was influenced by Charles Darwin's nineteenth-century theory of natural selection and subsequent research that has uncovered a 600,000-million-year fossil record and the mysteries of genetics. This was followed by Alfred Wegener's 1912 proposal of continental drift and the tectonic plate research that revealed planetary surfaces have been slowly moving "for at least the last 100 million years."[35] In concert with these three previous revolutions, climate research has, over the latter half of the twentieth century, displayed a dynamically balanced and yet cyclically changing climate that interrelates the planetary climate to geology, oceans, and evolving changes of biological life and humanity. These revolutions in scientific thought go beyond questions of earth physics, for they support Fleming's conclusion that each successive transformation in worldview has revealed the extent to which "history, climate, and culture are closely interwoven."[36] Or, as environmental historian J. R. McNeill states, the history of environmental issues clarifies the point that "history and ecology are, as fields of knowledge go, supremely integrative."[37]

IPCC UNCERTAINTIES

This change in Western research of human-climate relations has not been an easy shift, as is reflected in a debate that heated up in the 1970s. The warming trend that Keeling began observing from Mauna Loa continued until the early 1960s, but then in the 1970s, this pattern was thrown into disarray by a decade-long global cooling.[38] This decrease in temperature that lasted until the late-1970s drew international attention as some scientists projected a global cooling trend.[39] Other researchers who wanted to proceed slower with projections, because of the many uncertainties, appeared to be validated by the 1980s as the warming trend resumed. In the words of Ruddiman, "the few scientists who jumped to the premature conclusion that the minor climate cooling during the 1960s and 1970s was a harbinger of glaciation deserve the criticism they have received" because they overextended their analyses.[40] It was this debate that Weart argues led to an intensified "effort to understand how the climate system worked,"[41] resulting in the creation of global knowledge-sharing networks that culminated in the 1988 creation of the IPCC by the World Meteorological Organization and the United Nations. Utilizing the findings from thousands of researchers and the interests of governmental delegates representing more than a hundred countries, the IPCC has produced four reports in 1992, 1996, 2001, and 2007 that have successively found more evidence supporting a global warming trend related to human activities.

Based on this interdisciplinary and intergovernmental research, the IPCC has come to describe the climate's intricate complexity as being composed "of the atmosphere, the ocean, the ice and snow cover, the land surface and its features, the many mutual interactions between them, and the large variety of physical, chemical and biological processes taking place in and among these components."[42] Even with the IPCC's increasing understanding of the climate system and consensus that human systems are involved in the current changes, Weart has pointed out there remains significant uncertainties that can be seen as fuelling not only the 1970s warming-cooling debate, but also the continuing disagreement on what an adequate response will entail. Many of these uncertainties are based on a lack of knowledge concerning the possible nonlinear effects of rising greenhouse gases, or what are referred to as positive and negative feedbacks. Positive feedbacks refer to those complex interactions that can amplify the warming trend on local and global scales. In the North, these include interactions like the release of methane trapped in permafrost as the polar regions warm, and changes in surface reflectivity as snow and ice give away to vegetation.[43] Potentially mitigating the uncertainty of these interactions are negative feedbacks that can moderate changes through interactions like the warming-inspired increase of cloud cover and air moisture that will reflect more of the sun's energy away from the earth.

One significant way the IPCC's climate research has come to think about these complex feedbacks is through computer models that include everything from land cover and snow to gas chemistry and atmospheric pressure. As geographer David Demeritt explains, these models "simulate the behaviour of the climate system by dividing the earth into a three-dimensional grid and using supercomputers to solve mathematical equations representing exchanges of matter and energy between the grid points."[44] His concern with this approach is that where a lack of knowledge persists the variables tend to be hypothesized based upon the assumptions of the modeller. For example, Weart explains that these models often assume smooth and gradual climate changes despite paleoclimate evidence that suggests the interacting feedbacks have the potential to result in quick and abrupt changes.[45] This approach is often employed to maintain speed and power in computer projections. In a related critique, Demeritt states modellers tend to concentrate on simulating the most likely outcomes based on "a subjective judgment about risk tolerance" and the potential of system variability.[46] So while paleoclimate research of past feedbacks indicates that "atmospheric regimes can change within a few years and that large-scale hemispheric changes can evolve as fast as a few decades,"[47] the IPCC has tended to focus on computer models that project such changes to be unlikely during this century. That said, with each successive report, the IPCC has found these destructive potentialities to be more likely than previously projected.

To appreciate the revolutionary difficulties the IPCC faces in projecting the human role in the climate's complex feedbacks, it is helpful to consider a relatively recent debate on Gaia, climate change, and uncertainty that occurred in the journal *Climatic Change*. In 2002 and 2003 this journal hosted a number of articles by prominent biospheric theoreticians on the value of Gaia theory for elucidating principles about the complex climate system. Looking at the papers by Axel Kleidon, Timothy Lenton, and James Kirchner, Tyler Volk offered a commentary that defined what he saw "as tasks for the future of Gaia theory."[48] Though sympathetic to aspects of all three papers, he was in most agreement with Kirchner's assessment that the evidence of feedbacks that can impact the ecological conditions for life suggest biological interactions do not necessarily stabilize the planet. Considering the current increases of atmospheric CO_2, Kirchner states that documentation of only modest increases in carbon sequestration by vegetation acts as "an empirical rebuttal to Gaian notions of homeostasis and optimization" because it reveals atmospheric CO_2 is not regulated at an optimal biological set point.[49] While Kleidon and Lenton offer evidence for a more classic conception of Gaian self-regulation, Volk was of the opinion that "a few conforming examples drawn from the wealth of interactions within the biosphere" cannot be held up as evidence of general Gaian principles.[50]

In 2003 this debate continued with Lenton responding to the critiques of Kirchner and Volk by writing that "a combination of positive and negative feedback does not preclude regulation," as there are many examples of complex systems that regulate through a mixture of feedbacks.[51] While such a regulatory sense of Gaia may lead to visions of the rejected 1970s view of a global superorganism, he explains that Gaia theory today simply proposes four tenets: "life alters its environment," "life forms grow and reproduce," "the environment constrains life," and "natural selection occurs."[52] It is a view encapsulated by Primavesi when she describes Gaia as an entity "which possesses features of organization analogous to (not identical with) the physiological processes of individual organisms."[53] Responding to this nuanced view on Gaian regulation, Kirchner's 2003 article agreed that the coupling of atmosphere and biosphere should result in feedbacks and emergent behaviours like self-regulation, but he worried that such a view of Gaian self-regulation could lead to the problematic assumption that researching planetary systems is easy.[54] A check against research complacency was needed, in his opinion, because the progress in developing climate knowledge "is still dwarfed by what we don't know."[55]

While Weart argues that the persistent uncertainty underlying the IPCC's ever-increasing knowledge requires better interdisciplinary integration of the physical and social sciences,[56] Kirchner's approach promotes mechanistic thinking in a world where natural selection favours life forms that most effectively exploit "the environmental services" of their ecology.[57] This argument for mechanistic science is also very much grounded in the economizing mentality that Hallman, Sachs, and Primavesi are deeply concerned about. If this view was limited to the seemingly esoteric queries of Gaian theorists it would not be all that concerning, but it seems to also partake in a broader cultural trend. This concern is superficially observable in the title of the IPCC's chapter on the impact of climate change on animals, plants, and broader ecological processes: "Ecosystems and their Goods and Services."[58] Seemingly confirming Sachs's reference to a post-Brundtland sustainable management tendency, this title presumes it is possible to manage animals like polar bears as objectified "goods" and larger ecological processes like a melting Northwest Passage as "services" primarily oriented to human use. Such an approach is not isolated to the IPCC, for the 2004 interdisciplinary research represented in the *Arctic Climate Impact Assessment* states that northern changes will impact "local people and other living things that depend on these systems for food, habitat, and other goods and services."[59] This preoccupation with blending climate research and economic thought as a means for sustainably developing growth in a time of change can be conceived as the most trenchant uncertainty, for it is this cultural assumption that can be conceived as limiting the continued evolution of interdisciplinary research and, consequently, political response to Gaia's climatic changes and *Sila*'s northern warming.

The proposal by Kirchner that climate research must be more mechanistically grounded to assess a planet of self-interested economizing beings is in direct contrast to the "change of mind" Primavesi sees arising from the evolution of the IPCC's climate research and Gaian science. Because we have abused the earth systems "for our own benefit, with that benefit being computed, for the most part, in terms of monetary profit," she argues that such a belief system needs to be called into question.[60] As with Hallman and Sachs, Primavesi is concerned with how a mechanistic and economizing mind partakes in globalizing Western assumptions that are interrelated to contemporary environmental issues like climate change. In her words, the danger of our current over-reliance on science is "that it has presupposed our entitlement to handle global resources as though they are nothing but an inexhaustible supply of material for gratifying our desires."[61] More than that, this trust in an outdated version of science is based on the unfounded assumption that we can competently manage the planet in a way that will not impact future generations of humanity and biological life. Apparently, climate research needs to foster some capacity akin to the contextualizing power of Inuit *Silatuniq*. To further clarify the importance of this revolution in climate research, I will now look at another critique of the IPCC's approach that seemed to inform Canada's Conservative government's denial of climate change in relation to polar bears and a melting Northwest Passage.

CONSERVATIVE FRIENDS OF SCIENCE

As Canada's January 23, 2006, Election Day approached and it became increasingly clear the Conservatives were going to take power, environmentalists began expressing concern that a Stephen Harper government "would move Canada more into the same camp as U.S. President George W. Bush."[62] The potential implications to climate research of such an alignment was highlighted in a 2004 report by the Union of Concerned Scientists that found "the manipulation, suppression, and misrepresentation of science by the Bush administration is unprecedented," pointing to evidence that it actively discredited the IPCC's climate research, watered down the domestic 2003 *State of the Environment* report, and limited the scope and validity of the *Arctic Climate Impact Assessment*.[63] In Canada a similar critique of climate research was raised a few months following the Conservative election in a national newspaper that published an open letter to the prime minister signed by sixty researchers from the Friends of Science group.[64] Their approach to climate research is, as will be seen, relevant for thinking about Prime Minister Harper's view on Gaia's climate changes and *Sila*'s northern warming.

The Friends of Science letter expressed concern about the way in which inconclusive climate research had informed the previous Liberal government's embrace of the Kyoto Protocol and related policies they said squandered billions of dollars. Drawing upon the climate's complex uncertainties, these signatories not only argued that it "may be years yet before we properly understand the Earth's climate system" but that if the international community knew in the mid-1990s what is known now, Kyoto would have been seen as unnecessary.[65] Referencing the 1970s climate science debate, they reminded the prime minister and Canadians that only thirty years previous "many of today's global warming alarmists were telling us that the world was in the midst of a global-cooling catastrophe."[66] This message about scientific uncertainty was repeated at a May meeting between Harper and Frank Luntz, the former Republican climate advisor who advised President George W. Bush "to cultivate uncertainty when talking about climate change."[67] While the Republicans used public uncertainty to confront climate change and its research, Canada's previous three terms of Liberal government combined with a Conservative minority meant that, even if Harper wanted to, mainstream science could not be totally dismissed. In this political context, the cultivation of public uncertainty concerning climate research fell on the shoulders of nongovernmental organizations like the Friends of Science. They had a lot of public uncertainty to work with, as reflected in a 2006 Ipsos-Reid poll that found four out of ten Canadians related climate change "to natural warming and cooling patterns rather than human influences."[68]

Sensing the post-election opportunity needed to further the economic rationalization of Canada's climate research and policy, the Friends of Science conducted a spring campaign that included the open letter and the circulation of a *Climate Catastrophe Cancelled* DVD to politicians and the media.[69] In the year that followed, Canada's climate change response shifted significantly, though Albert Jacobs—a geologist and retired oil-exploration manager involved with the Friends of Science—admitted their "success is tied to the Conservative government."[70] The opening words of the October 19, 2006 press release for Harper's Clean Air Act highlighted this success: "After 13 years of Liberal inaction on the environment, Canada's new Government is getting things done to improve air quality and protect the health of Canadians."[71] Though the new act would legislate emission targets, fixed caps for greenhouse gases by 2025 and absolute reductions of 45 to 65 percent from 2003 levels by 2050,[72] its agreement with the Friends of Science position was reflected in the complete absence of any commitment to the Kyoto Protocol.

With Kyoto's emission mitigation policies sidelined, the government's proposed absolute reductions would be achieved through technology improvements in areas like CO_2 sequestration that it argued could reduce emissions by as much as 60 percent by 2050.[73] They also promoted an intensity-based approach that would reduce "the amount of greenhouse gases created by each unit of economic output," though environmental researchers and organizations argued it would not mean "Canada's total output of greenhouse gases would decline, as envisioned by Kyoto."[74] Rather, emissions would most probably rise due to trends like increasing consumption levels and fossil fuel production in the expanding Tar Sands.[75] In much the same spirit as Derocher's Polar Bears International critique of the government at the roundtable, the David Suzuki Foundation concluded that "the Act lacks meaningful targets, sets most timelines in the distant future, and focuses on emissions intensity—all of which guarantee continued rising pollution levels."[76] The consistency of this approach with the Friends of Science's economizing logic was further highlighted by the response of business leaders who were largely happy about these "clear, achievable targets."[77] It was particularly welcomed by the oil industry, which held the view that it was too difficult "to reduce total greenhouse-gas emissions while achieving the planned growth rates of oil sands output."[78]

It is important to highlight that the Friends of Science group is significantly different from other climate research networks like the IPCC in that it largely consists of oil geologists, Conservative insiders, and oil industry professionals.[79] Of the sixty scientists who signed the open letter, only one-third were Canadian and many were economists and geologists whose research had "received money from the oil, gas and coal industries in the United States."[80] One of the signatories who was promoted by the Friends of Science in its spring media campaign following Harper's election was the Canadian climate scientist Dr. Timothy Ball.[81] In his cross-country speaking tour to federal and provincial Conservative representatives, as well as newspaper editorial boards, Ball argued "Environment Canada and other agencies fabricated the climate change scare in order to attract funding."[82] Interestingly, his presentations neither revealed the Friends of Science as a funding source nor mentioned he had not published a climate research paper in a decade and a half.[83] It is a point consistent with Weart's finding that the research often touted by critics of the IPCC is generally so limited as to be unacceptable to "peer-reviewed journals where every statement was reviewed by other scientists."[84] This is not to suggest that all involved in the Friends of Science are unpublished. During the Liberal government's tenure, two signatories, applied mathematician Christopher Essex and economist Ross McKitrick, wrote a popular nonacademic Canadian book on climate change that argued the IPCC and Environment Canada were creating a "convection of certainty" to support ratifying the Kyoto Protocol.[85] While conceding it may eventually be proven that humans are affecting the climate, they conclude, as with the open letter and Ball, that climate research is being undermined by its politicization and that this represents a "loss of nerve" in economic rationality.[86]

The extent to which this critical conservative view on climate research's uncertainties is valid can be assessed by returning to the debate on Gaia in the journal *Climatic Change* and the letter of response by the theory's originator James Lovelock. Exemplifying his capacity for, as Volk put it, "asking big questions," Lovelock compared Gaia's climate complexity to quantum uncertainties. There was for him an important similarity between a quantum theory that "is incomprehensible because the universe itself is far stranger than the human mind can contemplate," and a Gaia theory that "is difficult to understand because we are not used to thinking about the Earth as a whole system."[87] The validity of making such a conceptual analogy is supported by the research of physicist Ian Barbour on the interdisciplinary implications of quantum uncertainty. He begins by defining the "Heisenberg uncertainty principle" as a quantum reality that hinders researchers from simultaneously knowing the position and velocity of an electron due to the perplexing way in which knowledge of one variable increases as the other decreases.[88] This view of uncertainty is the latest scientific conception on the nature of uncertainty, with Barbour identifying two forerunners that are important to understand because they co-exist in the above climate change debates, both as valid partial views on phenomena and as limiting scientific predispositions.

In the same spirit as Fourier's nineteenth-century search for a "grand geophysical law" of the climate, the first approach defined by Barbour attributes uncertainty "to temporary human ignorance, in the conviction that there are exact laws which will eventually be discovered."[89] Complete knowledge of climate complexity is possible in this view once the basic physical or mechanistic laws are clarified. Uncertainty in the second approach is seen as being due to the limits of conceiving and researching a reality that "the observer inevitably disturbs."[90] From this perspective, it could be said that researcher uncertainty concerning climate change is related to the way in which human and cultural emissions undermine complete objectivity. It is the latest quantum approach of defining uncertainty as nature's indeterminacy that Barbour finds to be the most inclusive because it understands the other two uncertainties—scientific ignorance and epistemological limitations—as being primarily related to fundamental properties of nature that transcend complete knowledge.[91] Summarizing this situation, Barbour writes that the "future is not simply unknown, it is not decided," though it is not completely open because the future's possibilities depend upon present actions.[92] Based on this definition of uncertainty, one can see why Lovelock made the connection he did, for the uncertainty surrounding climate change can likewise be seen as a function of Gaia's indeterminate response to those cultural values that inform today's rising emissions, the research of these issues, and the arising political economic debates.

Responding to Kirchner's call for more mechanistic research, Lovelock argues that just because reductionist science cannot rationalize quantum uncertainties or the emergent properties of Gaian systems "does not mean that these phenomena do not exist."[93] Such an analysis suggests the approach to climate change promoted by the Friends of Science, and seemingly taken up by Canada's Conservative government, is based in an outdated mechanistic view of uncertainty that focuses on delaying action until complete knowledge is obtained. For researchers like Kirchner, Essex, McKitrick, Ball, and others who are finding it difficult to make such a shift in thought, Lovelock advises them to leave Cartesian scruples aside and consider that "you do not need to know the details of a friend's biochemistry to know them as a person and in a similar way you can envisage Gaia without knowing the recondite details of its geochemistry."[94]

The thought of Primavesi goes one step further by directly confronting such economizing assumptions that seemingly hinder both the Friends of Science position and the IPCC's interdisciplinary research. She argues that the individual self-interest at the core of this economizing mentality educates us to ignore what climate change is today teaching: "that without earth's givenness built up over deep time, and without the present gifts of life-support systems made possible by it, we would not exist, could not exist."[95] According to Primavesi, a paradigm shift in our thinking about climate change is being asked of us by the regulatory patterns of the Gaian system. As with the critiques of Hallman and Sachs, Gaia's indeterminism calls for more balanced interdisciplinary climate research that can wisely contextualize mechanizing and economizing assumptions. This goes far beyond the IPCC, and is light years beyond a conservative Friends of Science approach that denies the need for a climate response until some never-arriving future when certain knowledge has been attained.

CONCLUSION: TOWARD RESEARCH *SILATUNIQ*

The methodological response to indeterminism in quantum physics has been a complementarity that brings together various perspectives on phenomena in question. Explaining the reasoning for this approach, Barbour writes that "differing aspects of the structure of events are interpretable by differing models, each of which is incomplete and applicable only to certain experimental situations."[96] Consequently, knowledge from multiple viewpoints increases the understanding of the indeterminate phenomena, though uncertainty is never completely reduced. This sounds remarkably similar to the IPCC interdisciplinarity that is finding, as Kirchner states, its ever-increasing knowledge is being continually "dwarfed by what we don't know." Such a complementarity conflicts with the approach proposed by the Friends of Science and, more specifically, Essex and McKitrick, who not only have problems with the IPCC's interdisciplinarity but also argue that climate research informed by Inuit perspectives has limited validity. Specifically referring to the research with the Inuit of Sachs Harbor that led to the documentary, *Sila Alangotok*, they write that connecting "a non-random sample of Inuit recollections about recent springtime temperatures" to rising greenhouse gases and northern warming is highly suspect.[97] If the Conservative silence on climate change at the Polar Bear Roundtable represented its continued agreement with the Friends of Science view, then their roundtable agreement with Inuit to better incorporate IQ in the conservation of polar bear populations either reflected a divergence from this position or a more palatable political move than being aligned with climate research. While political motives will continue to be a concern in the chapters that follow, my primary focus here has been to clarify those cultural assumptions that are limiting a complementarity of views in climate research.

Grappling with the relevance of Gaian indeterminism to climate research and policy is central to Primavesi's thought, for as a theologian she recognizes that "unknowability is not to be confused with mere ignorance" when talking about realities that transcend knowledge.[98] In Barbour's analysis on the indeterminate limits to knowledge, he similarly concludes that when science, the humanities, and religion become aware of the complementarity between their different types of knowledge, then it may be possible to develop a methodology where rationality "is fulfilled, not abrogated, by revelation."[99] Extending this view to environmental sustainability, Barbour further proposes that a viable future may depend upon scientists sharing their knowledge of ecological changes with religious leaders who can contemplate the meaning of these changes to ways of living.[100] Rather than stepping back from an interdisciplinary and intercultural complementarity in climate research, as argued by the Friends of Science, there is seemingly a need to be more inclusive in the range of voices that can influence both research and policy. Considering this Gaian change of mind, Primavesi asks whether the West's increasing, though limited, knowledge of climate change will make us better "able, willing and competent to act wisely"?[101] Such a query seems to call forth the image of the Inuit shaman Najagneq, suffering out in the great loneliness as his IQ was initiated into a contextualizing *Silatuniq*, except here the initiated is Western climate research and the initiator is Gaia's climatic response to the political economics of rising greenhouse gas emissions.

This need for climate research to be initiated into *Silatuniq* was made apparent to the Inuit Circumpolar Conference—now referred to as the Inuit Circumpolar Council—as it participated with scientific, governmental, and indigenous organizations in completing the 2004 *Arctic Climate Impact Assessment*. Though the IPCC and *Arctic Climate Impact Assessment* offer similar northern climate projections based on comparable networks of international and interdisciplinary researchers, the latter endeavour went one step further by incorporating the knowledge of six indigenous cultures from the global North: the Aleut International Association, Arctic Athabascan Council, Gwich'in Council International, Russian Association of Indigenous Peoples of the North, the Saami Council, and the Inuit Circumpolar Conference.[102] Despite this indigenous involvement, a few months after the *Arctic Climate Impact Assessment*'s release, Sheila Watt-Cloutier—then chair of the Inuit Circumpolar Conference—co-authored a response that clarified the limited nature of a report that "does not address targets or timetables for reduction of greenhouse gas emissions."[103] She related that the six indigenous organizations "submitted a statement to be included at the beginning of the science assessment and summary volumes," and it was "rejected as being too political."[104] One excerpt states:

> To Arctic Indigenous peoples climate change is a cultural issue. We have survived in a harsh environment for thousands of years by listening to its cadence and adjusting to its rhythms. We are part of the environment and if, as a result

of global climate change, the species of animals upon which we depend are greatly reduced in number or location or even disappear, we, as peoples would also become endangered.[105]

While the above statement was too political for a research project whose participants and interests are largely based in the south, the Inuit Circumpolar Conference argued a more forceful statement was needed in light of the potential climate impact on northern ecologies, knowledge, and cultural ways. This critique supports the suggestion that climate research's interdisciplinary and intercultural scope is being restricted by scientific and economic assumptions. As I will argue over the next three chapters, this skewing influence has a long Canadian, American, and Western history that is limiting our wise research of Gaia's indeterminate climate changes and *Sila*'s northern warming.

ENDNOTES

1   Sierra Club of Canada, "Harper's Position on Kyoto: A Tragedy for the Planet," 12 January 2006, http://www.sierraclub.ca.

2   Polar Bears International, "National Roundtable on Polar Bears, Media Release," 16 January 2009, http://www.polar-bearsinternational.org/in-the-news/polar-bear-roundtable/.

3   David G. Hallman, "Climate Change: Ethics, Justice, and Sustainable Community," in D. T. Hessel and R. R. Ruether, eds., *Christianity and Ecology: Seeking Well-being of Earth and Humans* (Cambridge: Harvard University Press, 2000), 453–471.

4   Ibid., 457.

5   Wolfgang Sachs, "Environment," in W. Sachs, ed., *The Development Dictionary* (London: Zed Books 1992), 27.

6   Ibid., 32.

7   Spencer Weart, *The Discovery of Global Warming* (Cambridge: Harvard University Press, 2003), 198.

8   Anne Primavesi, *Gaia and Climate Change: A Theology of Gift Events* (London: Routledge/Taylor & Francis Group, 2009), 10. For a complementary eco-theological analysis of climate change, see Michael S. Northcott, *A Moral Climate: The Ethics of Global Warming* (London: Darton, Longman and Todd, 2007).

9   Primavesi, *Gaia and Climate Change*, 126.

10   Weart, *The Discovery of Global Warming*; Gale E. Christianson, *Greenhouse: The 200-Year Story of Global Warming* (New York: Walker and Company, 1999); James Rodger Fleming, *Historical Perspectives on Climate Change* (New York: Oxford University Press, 1998).

11   Christianson, *Greenhouse*, 115.

12   Quoted in Christianson, *Greenhouse*, 115.

13   John Robert McNeill, *Something New under the Sun: An Environmental History of the Twentieth-Century World* (New York: W. W. Norton and Company, 2000), 14.

14   Thomas F. Homer-Dixon, *The Ingenuity Gap* (Toronto: Vintage Canada, 2001).

15   IPCC, *Climate Change 2007: The Physical Science Basis, Summary for Policymakers, Contribution of Working Group I to the Fourth Assessment Report of the IPCC* (Geneva, Switzerland: IPCC Secretariat/World Meteorological Organization, 2007), 2, 4.

16   Fleming, *Historical Perspectives on Climate Change.*

17   Ibid., 21.

18   Quoted in Fleming, *Historical Perspectives on Climate Change*, 18.

19   Ibid.

20   Ibid., 24.

21   Ian G. Barbour, *Issues in Science and Religion* (New York: Harper and Row, 1971), 57.

22   Julie Cruikshank, *Do Glaciers Listen? Local Knowledge, Colonial Encounters, and Social Imagination* (Vancouver: UBC Press, 2005), 143.

23   Ibid., 149, 243.

24   Ibid., 32, 243.

25   Quoted in Cruikshank, *Do Glaciers Listen?* 139.

26   Ibid.

27   Fleming, *Historical Perspectives on Climate Change*, 47.

28   Doug Macdougall, *Frozen Earth: The Once and Future Story of Ice Ages* (Berkeley: University of California Press, 2004).

29   Ibid.

30   Quoted in Fleming, *Historical Perspectives on Climate Change*, 63.

31   Ibid., 82.

32   Ibid.

33   W. F. Ruddiman, *Plows, Plagues, and Petroleum: How Humans Took Control of Climate* (Princeton: Princeton University Press, 2005).

34   Ibid., 8.

35   Ibid.

36   Fleming, *Historical Perspectives on Climate Change*, 136.

37   McNeill, *Something New under the Sun*, 362.

38   Weart, *The Discovery of Global Warming*; Fleming, *Historical Perspectives on Climate Change*.

39   Ruddiman, *Plows, Plagues, and Petroleum*; Weart, *The Discovery of Global Warming*; Fleming, *Historical Perspectives on Climate Change*.

40   Ruddiman, *Plows, Plagues, and Petroleum*, 105.

41   Weart, *The Discovery of Global Warming*, 94.

42   A. P. M. Baede et al., "The Climate System: An Overview," in contribution of Working Group I to the Third Assessment Report of the Intergovernmental Panel on Climate Change, *Climate Change 2001: The Scientific Basis* (Cambridge: Cambridge University Press, 2001), 87.

43   Susan Joy Hassol et al., *Impacts of a Warming Arctic: Arctic Climate Impact Assessment (ACIA)* (Cambridge: Cambridge University Press, 2004), 34.

44   David Demeritt, "The Construction of Global Warming and the Politics of Science," *Annals of the Association of American Geographers* 91, no. 2 (2001): 325.

45   Weart, *The Discovery of Global Warming*.

46   Demeritt, "The Construction of Global Warming and the Politics of Science," 325.

47   IPCC, *Climate change 2001: Synthesis Report: Contributions of Working Group I, II, and III to the Third Assessment Report of the IPCC* (Geneva: IPCC Secretariat/World Meteorological Organization, 2001), 53.

48   Tyler Volk, "Toward a Future for Gaia Theory: An Editorial Comment," *Climatic Change* 52 (2002): 423–430. He reviews the following articles: J. W. Kirchner, "The Gaia Hypothesis: Facts, Theory, and Wishful Thinking," *Climatic Change* 52 (2002): 391–408; A. Kleidon, "Testing the Effect of Life on Earth's Functioning: How Gaia Is the Earth System?" *Climatic Change* 52 (2002): 383–389; T. Lenton, "Testing Gaia: The Effect

of Life on Earth's Habitability and Regulation," *Climatic Change* 52 (2002): 409–422.

49   Kirchner, "The Gaia Hypothesis: Facts, Theory, and Wishful Thinking," 405.

50   Volk, "Toward a Future for Gaia Theory," 424.

51   T. Lenton and D. Wilkinson, "Developing the Gaia Theory: A Response to the Criticisms of Kirchner and Volk," *Climatic Change* 58 (2003): 4.

52   Ibid., 3.

53   Anne Primavesi, *Sacred Gaia: Holistic Theology and Earth System Science* (London: Routledge, 2000), 5.

54   J. W. Kirchner, "The Gaia Hypothesis: Conjectures and Refutations," *Climatic Change* 58 (2003): 22, 42.

55   Ibid., 42.

56   Weart, *The Discovery of Global Warming*, ix.

57   Kirchner, "The Gaia Hypothesis: Facts, Theory, and Wishful Thinking," 399.

58   Habiba Gitay et al., "Ecosystems and Their Goods and Services," in contribution of Working Group II to the TAR of the IPCC, *Climate Change 2001: Impacts, Adaptation and Vulnerability* (Cambridge: Cambridge University Press 2001).

59   Hassol et al., *Impacts of a Warming Arctic: Arctic Climate Impact Assessment* (ACIA), 45.

60   Primavesi, *Gaia and Climate Change*, 130.

61   Ibid., 10–11.

62   Sierra Club of Canada, "Canadian Climate Coalition Denounces Conservative Party for Ducking the Issues," 17 January 2006, http://www.sierraclub.ca.

63   See Robert F. Kennedy, *Crimes Against Nature: How George W. Bush and His Corporate Pals Are Plundering the Country and High-Jacking Our Democracy* (New York: HarperCollins, 2004), 95.

64   *National Post*, "Open Kyoto to Debate: Sixty Scientists Call on Harper to Revisit the Science of Global Warming," 6 April

2006, http://www.canada.com/nationalpost/financialpost/
story.html?id=3711460e-bd5a-475d-a6be-4db87559d605.

65 Ibid.

66 Ibid.

67 Charles Montgomery, "Nurturing Doubt about Climate
Change Is Big Business," *Globe and Mail*, 12 August 2006, sec. F,
p. 4.

68 Quoted in Montgomery, "Nurturing Doubt about Climate
Change Is Big Business," sec. F, p. 4.

69 Ibid.

70 Montgomery, "Nurturing Doubt about Climate Change Is
Big Business," sec. F, p. 5.

71 Rona Ambrose, Environment Minister, "Clean Air for All
Canadians," 19 October 2006, http://www.conservative.ca/
EN/2459/56332.

72 Ibid.

73 Bill Curry and Mark Hume, "PM Plans 'Intensity' Alternative to Kyoto: Blueprint Wouldn't Necessarily Reduce Emissions; Critics React with Scorn," *Globe and Mail*, 11 October
2006, sec. A, p. 1.

74 Martin Mittelstaedt, "What Does Ottawa's Green Plan
Entail?" *Globe and Mail*, 20 October 2006, sec. A, p. 5.

75 Curry and Hume, "PM Plans 'Intensity' Alternative to
Kyoto."

76 David Suzuki Foundation, "New Federal Clean Air Act
Won't Clean Air," 19 October 2006, http://www.davidsuzuki.
org/climate_change.

77 Bill Curry, "Critics Blast Ottawa's 'Shameful' Green Plan,"
Globe and Mail, 20 October 2006, sec. A, pp. 1, 4.

78 Shawn McCarthy, "A Red Flag in the Global-Warming
Battle," *Globe and Mail*, 11 October 2006, sec. A, p. 4.

79 Quoted in Montgomery, "Nurturing Doubt about Climate
Change Is Big Business," sec. F, p. 4.

80 Ibid., sec. F, p. 5.

81 Ibid., sec. F, p. 4.

82 Ibid., sec. F, pp. 4–5.

83 Ibid., sec. F, p. 4.

84 Weart, *The Discovery of Global Warming*, 166.

85 Christopher Essex and Ross McKitrick, *Taken by Storm: The Troubled Science, Policy and Politics of Global Warming* (Toronto: Key Porter Books, 2002), 231.

86 Ross McKitrick, *An Economist's Perspective on Climate Change and the Kyoto Protocol*, (presentation to the Department of Economics Annual Fall Workshop, University of Manitoba, Winnipeg, MA, 7 November 2003); Essex and McKitrick, *Taken by Storm*.

87 James E. Lovelock, "Gaia and Emergence: A Response to Kirchner and Volk," *Climatic Change* 57 (2003): 1.

88 Barbour, *Issues in Science and Religion*.

89 Ibid., 298.

90 Ibid.

91 Barbour, *Issues in Science and Religion*, 299; Shimon Malin, *Nature Loves to Hide: Quantum Physics and Reality, a Western Perspective* (Oxford: Oxford University Press, 2001).

92 Barbour, *Issues in Science and Religion*, 304–305.

93 Lovelock, "Gaia and Emergence," 1–2.

94 Ibid., 3.

95 Anne Primavesi, *Gaia's Gift: Earth, Ourselves and God After Copernicus* (London: Routledge/Taylor & Francis Group, 2003), 135.

96 Barbour, *Issues in Science and Religion*, 290–291.

97 Essex and McKitrick, *Taken by Storm*, 54–55.

98 Primavesi, *Gaia's Gift*, 45.

99 Barbour, *Issues in Science and Religion*, 268.

100 Ian Barbour, "Scientific and Religious Perspectives on Sustainability," in D. T. Hessel and R. R. Ruether, eds., *Christianity and Ecology: Seeking the Well-being of Earth and Humans* (Cambridge: Harvard University Press, 2000).

101 Primavesi, *Gaia and Climate Change*, 73.

102 Sheila Watt-Cloutier, Terry Fenge, and Paul Crowley, *Responding to Global Climate Change: The Perspective of the Inuit Circumpolar Conference on the Arctic Climate Impact Assessment*, 16 February 2005, http://www.inuitcircumpolar.com.
103 Ibid.
104 Ibid.
105 Ibid.

CANADIAN CALL FOR A GLOBAL CONSCIENCE

At the December 2005 Montreal Conference of the United Nations Framework Convention on Climate Change, Canada's then Liberal Prime Minister Paul Martin addressed the gathering by stating climate change is "a global challenge that demands a global response, yet there are nations that resist, voices that attempt to diminish the urgency or dismiss the science."[1] With 190 nations participating and approximately ten thousand international delegates representing governments, environmental organizations, business leaders, and cultural interests, it was at that time the largest climate change gathering since the 1997 Kyoto meetings. They were there to negotiate commitments to greenhouse gas emission reductions beyond the Kyoto Protocol end date of 2012, and to get resistant nations involved in the process. The Canadian host went on to tell the delegates that "time is past to seek comfort in denial," for it is no longer possible "to pretend that any nation can stand alone, isolated from the global community."[2] Later with the press, Prime Minister Martin specifically called upon the reticent United States to listen to the international community and respond with a "global conscience."[3] Though this call for a political response to the Intergovernmental Panel on Climate Change's (IPCC's) findings was directed at American President George W. Bush's denial of climate policy and research, it was also a veiled critique of Stephen Harper's Conservative opposition view at the beginning of a national election campaign. The irony of this Liberal call for a "global conscience" was that, leading up to the conference, Canada's emissions had increased by 24 percent in comparison with the United States' 13 percent, despite Canada having ratified the Kyoto Protocol.

This poor enactment of Canadian climate policy was primarily a Liberal failure, for they were in power since 1993 and the governing party for about thirty of the previous forty years. There was a partial acknowledgement of this issue in Prime Minister Martin's address when he spoke of Canadian energy production and consumption trends that made "our record on combating climate change far from perfect."[4] One mitigating factor in this failure was the North American Free Trade Agreement, which economically and energetically connected Canada with an American nation that had, in 2001, changed from Democratic to Republican policies, and was thus putting economic pressure on the Liberals to harmonize their climate policies.[5] While Canada actively pursued international leadership in sustainability, Anthony Perl and Eugene Lee explain the nation actually continued reaping the "rewards of ever-closer economic integration with the United States, a country that increasingly behaves as if global sustainability was someone else's problem."[6] The resulting contradiction made it impossible for the Liberal government to even approach its Kyoto Protocol commitments, not to mention being given an international environmental ranking "second worst among twenty-eight Organization for Economic Cooperation and Development member nations, followed only by the United States."[7] It was in this continental context of competing liberal and conservative visions that Prime Minister Martin was in Montreal following the traditional Canadian approach of using international institutions to buffer the powerful political influence of their neighbour[8]—and thus hoping to lessen the contradiction between Liberal policy goals and performance in responding to Gaia's indeterminate climate changes.

An intimation of the global conscience proposed by Prime Minister Martin was highlighted when he spoke of the developed world creating a situation where developing nations "will suffer most if the effects of climate change set off an even worse decline in local living conditions."[9] In a Canadian context, this is exactly the unjust situation that the Inuit Circumpolar Conference has been responding to on behalf of Inuit who are facing a warming *Sila*, melting Northwest Passage, and descent of polar bears upon their communities. Ever since its founding in 1977, the Inuit Circumpolar Conference has advocated for a greater inclusion of *Inuit Qaujimatuqangit* (IQ) and Inuit interests in northern "strategies for environmental management and sustainable development."[10] Though the focus of climate research on traditional ecological knowledge often marginalizes IQ's cultural and spiritual views, neither the IPCC nor the *Arctic Climate Impact Assessment*, referred to in earlier chapters, propose a complete rejection of these considerations that are important to the Inuit Circumpolar Conference. Rather, the IPCC clarifies that cultural, religious, and aesthetic values need to be accounted for, but because they are difficult to measure, a monetary value can be assessed through proxies like "a tourist's willingness to pay to see wildlife in natural habitats."[11] This approach may be useful, for example, to Inuit who are concerned with maintaining the economic benefits of hunting polar bears, but it also clearly extends the cultural uncertainty of a wide-reaching economizing assumption. It was an issue that Jaypeetee Arnakak often expressed concern about, for he saw the West's dominant prescriptive approach of incorporating IQ's traditional ecological knowledge into environmental management, while neglecting anything approaching *Silatuniq*, as deeply problematic.[12] In this chapter, I will survey the role of these economizing trends in limiting Canada's political economic manifestation of a global or northern conscience.

A PROGRESSIVE FAILURE

Evoking the symbolism of Montreal as a place where the international community came together in 1987 to ban chemicals related to the ozone hole in the Montreal Protocol, Prime Minister Martin asked the delegates to work at making Montreal "synonymous with the moment the world came together, and together set off down the long but vital path to progress, real progress, progress we can measure, progress we can one day celebrate."[13] The irony was that the Montreal Protocol evoked by Martin was not hosted by the Liberals but rather occurred during the nine-year interlude of a Progressive Conservative government that significantly impacted subsequent Liberal climate policy. Following the Montreal Protocol, this conservative government changed the approach of having scientists directly involved in international decision-making processes because, in Steven Bernstein's words, scientists "reflected too closely a concern with the environment rather than effective or pragmatic responses acceptable to government."[14] The December 1991 international climate negotiations was the first time Canadian "economic experts replaced senior atmospheric scientists and forestry experts."[15] This shift followed international trends of utilizing market economic approaches in environmental responses that Wolfgang Sachs, as discussed in the previous chapter, associates with the Brundtland Commission's elucidation of sustainable development in 1987.[16] At the 1992 Rio de Janeiro Earth Summit, the first international agreement on curbing greenhouse gas emissions was reached, though its largely voluntary nature meant there was little progress prior to the 1997 Kyoto Protocol. When Jean Chrétien returned the Liberals to power in 1993, with Paul Martin as his Finance Minister, the road to "real progress" continued to be primarily measured by an economic frame of reference that would, by the Montreal conference, be implicated in a Canadian failure of conscience.

In the lead up to the 1997 Kyoto, Japan meeting, the IPCC released its *Second Assessment Report* that concluded the evidence suggests "there is a discernible human influence on global climate."[17] As this rising scientific consensus increased the pressure for a political response that went beyond voluntary measures, there also grew an international divide between the European Union and the then Democratic-led United States, the latter of which wanted to promote less stringent emission reductions.[18] There was general agreement concerning the six gases—carbon dioxide, the three halocarbons, methane, and nitrous oxide—that needed emission reductions. However, problems arose around the question of how much and by whom. Canada's Liberal government attempted to negotiate the gap between America and Europe, and in the process reinvigorate its international image "as a facilitator of global agreement and compromise."[19] But with the Kyoto negotiations continuing to stall, it was Democratic Vice President Al Gore's arrival that led to the United States' acceptance of a 7 percent reduction of greenhouse gas emissions below 1990 levels by 2010, a higher than expected agreement that amounted to a 23 percent reduction based on emission increases since 1990.[20] Other developed countries signed on to similar agreements, with the European Union at 8 percent, and other industrialized nations at a minimum of 5.2 percent of 1990 levels. Because the American response surpassed Canada's initial commitment, the Liberal government felt pressured into a 6 percent reduction that amounted to a 19 percent decrease due to emission increases since 1990.[21] As in Montreal eight years later, Canada's Liberal environmental image was partially being measured in comparison to political economic forces south of its border.

An even more difficult issue for the Kyoto Protocol than the question of emission reductions was how to conscionably deal with the historic issues of equity between the world's rich and poor nations. Ever since the Brundtland Commission and Rio, the United Nations pushed for a climate response that would ensure current and "future generations will live under climate conditions that permit sustainable social and economic development."[22] While the history of development can be traced back to President Harry S. Truman's 1949 speech that "scientific advances and industrial progress" will be made "available for the improvement and growth of underdeveloped areas,"[23] Gustavo Esteva points out the term "development" had in fact replaced earlier worn-out colonial ideas like 'improvement' and 'progress.'[24] Development critics like Esteva and Gilbert Rist argue the latter half of the twentieth century saw many nations reclassified as "underdeveloped" and thus "called upon to deepen their Westernization by repudiating their own values."[25]

The continuation of various international economic inequities and devastating social issues were aggravated over the latter part of the century by arising local and global environmental issues. It was in this recurring historic context of developmental failures that in 1987 the United-Nations-sponsored Brundtland Commission defined the new era of sustainable development. As it states, "technology and social organization can be both managed and improved to make way for a new era of economic growth."[26] This recycling of development and progress is, according to Rist, a gross contradiction because the economic growth policy that is proposed to alleviate social and environmental issues in reality does not differ "from the policy which historically opened the gulf between rich and poor and placed the environment in danger."[27] It was in the context of these progressive Western visions that the Kyoto Protocol likewise proposed to sustain international economic growth while reducing emissions.

At the Kyoto meetings it took various political efforts to negotiate the "intractable problems involving fairness and the power relations between industrialized and developing countries."[28] Beyond the issue of Western nations being the dominant industrial source of past and present greenhouse gas emissions, these power relations were also problematic because, as David Hallman states, "global warming will affect the whole planet with the South suffering disproportionately."[29] Within Canada a similar dynamic reverses this North-South divide, for it is Inuit who are dealing with *Sila*'s northern warming and related polar bear changes due to emissions that nationally have originated mostly in the more populated and industrially developed regions to the south. The Kyoto Protocol proposed to solve these equity issues on an international scale by having developed countries like Canada, the United States, and the European Union agree to greenhouse gas reductions, while a list of developing countries would maintain or increase emissions so as to foster sustainable development. These commitments reflected the responsibility developed Western nations felt they had for the onset of climate change, as well as the continued need developing nations had for economic growth.

One of the biggest obstacles to successfully enacting the Kyoto Protocol's sustainable development vision occurred when George W. Bush was elected president of the United States in 2000 and soon thereafter announced the agreement would not be ratified because "regulation would inflict economic disaster on Americans" and it "would turn over the world economy to the unregulated developing countries."[30] Despite its economic ties with the United States, Canada ratified the Kyoto Protocol amidst diverse critiques concerning the viability of an agreement that did not include the world's largest greenhouse gas emitter. The Canadian Association of Petroleum Producers stated they had reservations as to "how the government will reconcile the targets agreed to at Kyoto and its commitment to no punitive taxes, more jobs, and continued economic growth."[31] This political situation was an environmental conundrum, for it set long-term and global interests against the "short-run budgetary and political impacts" needed for maintaining Liberal power.[32] As Perl and Lee explain, America's rejection of imposing costs and preference for market mechanisms in environmental protection "has never been far from Canadian economic consciousness," and as such national business interests lobbied for a slow policy response to Kyoto.[33] In response to sceptical business interests, Prime Minister Chrétien's government initiated a First National Climate Change Business Plan that targeted economic sectors for emission reductions through voluntary programs.[34] Domestic economic pressures continued to confine Canada's climate policy well into the first decade of the twenty-first century, with the Liberal government's focus epitomized by then Finance Minister Paul Martin's comments that "dealing with climate change provides opportunities for Canadian companies to make money and that developing new technology is key."[35] This harmonization of external political and internal economic pressures would significantly impede Canada's Liberal commitment to Kyoto.

After Paul Martin became prime minister in 2003, his government released a report entitled *Moving Forward on Climate Change*. While considering some of the factors involved in Canada's failure to approach its Kyoto commitments, it was noted the economy performed "better than had been projected" as the gross domestic product grew by 43 percent rather than the forecasted 34 percent.[36] This set of events fostered a revised plan that would "harness market forces" to transform its growth economy toward more sustainable energy consumption via a combination of emission reductions and renewable energy technology. During the Montreal conference, this idea of transforming the economy toward renewable energy received support in an editorial article written by the provincial premiers, Gary Doer of Manitoba and Jean Charest of Quebec. Envisioning economic opportunities in green-building technology, hydrogen, hybrid transportation, and energy efficiencies, Doer and Charest stated it was time "to innovate and encourage ideas, and build our economy in a way that brings prosperity, health and social well-being, and sustainability."[37] The new Liberal plan also projected these initiatives could contribute "to cleaner air for Canada's cities, enhance biodiversity, help to preserve wild spaces and generally improve the quality of life."[38]

A 2004 governmental report on some potential regional effects of climate change highlighted this economizing tendency that would ensure Martin's new "path to progress" could not support a global conscience. Discussing the overall approach of the report, it states that "the concepts and methods of economics have been recognized as a principal means of translating scientific research on climate change into policies."[39] Continuing the earlier Progressive Conservative shift from a primarily scientific view, it adds, "assessing the economic impacts of climate change involves estimating the value of direct and indirect market and nonmarket impacts, the costs of implementing adaptation options and the benefits gained as a result of the adaptation."[40] Such a cost-benefit approach has led many economists to argue that adapting to climate change is more feasible than the Kyoto Protocol's greenhouse gas mitigation because the latter will only provide small environmental benefits while imposing significant costs. They also conclude that adaptation measures like improved forecasting, sea walls, and emergency preparation programs can at least complement, and at best be an alternative to, mitigation strategies.

The problem is that some climate research projections suggest the uncertainty "is too large to apply cost-benefit analysis."[41] As the IPCC explains, there are normal climate variations that socio-economic systems can address, but it is with the extreme projections that adaptation as a long-term approach becomes questionable.[42] That said, the IPCC has tended to use models of a gradually changing climate that support a cost-benefit approach, even though, as Weart explains, common sense recognizes "when you push on something steadily it may remain in place for a while, then move with a jerk."[43] He further adds that this abrupt potential is largely overlooked by politicians, economists, the media, and the public. This seems to be the case in Canada where concerns about being harmonized with American policies led it to promote cost-benefit adaptation measures while hoping for a successful voluntary reduction of greenhouse gases. In contrast, Gaia's indeterminate changes and *Sila*'s northern warming suggest such economic analyses need to be contextualized by broader interdisciplinary and intercultural climate research.

As the Montreal conference approached, evidence indicated that some nations were performing well in approaching their emission reduction commitments. While the United Kingdom's 13 percent and Iceland's 8.2 percent reductions below 1990 levels highlighted the potential success of progressive national policies aimed at energy efficiency and renewable energy technology, most Western nations had emissions hovering around, or higher than, those of 1990. The overall assessment was that developed nations that had ratified the Kyoto Protocol had "reduced emissions by 5.9 percent compared to 1990 levels,"[44] but this limited success only further highlighted the fact that nonratifying nations like the United States were creating an economic and political drag on the progress of other closely aligned nations like Canada. Added to this picture of mixed international success was the fact that the reductions of many countries, like those of the former Soviet Union, were related to economic downturns rather than policies. The most significant decrease of emissions in the European Union was Germany's 18.2 percent, but a significant portion of that was related to the inclusion of East Germany's economic recession. In light of more recent IPCC projections that suggest just stabilizing the climate to a 2°C rise will require industrialized nations to cut emissions by 80 to 90 percent from 1990 levels by 2050,[45] it may be that Canada's Liberal approach was merely an extreme failure in a progressive tradition that is incompatible with a global or northern conscience.

Despite the United States' resistance to a climate response, Prime Minister Martin's call for equating Montreal with progress had some initial success when the conference ended on December 10, 2005 with the Montreal Action Plan. The deal outlined compliance rules, agreed upon a schedule for developed countries to negotiate emission reductions for a period following Kyoto, and promoted increasing carbon dioxide sequestration and capture technologies. What proved to be less successful for Prime Minister Martin was the Liberal strategy of linking Stephen Harper's Conservatives with America's Republicans. With two weeks left in the election campaign and polls indicating the Liberal government was about to be defeated, Martin's critique was supported by a number of environmentalists. The executive director of the Sierra Club, Elizabeth May, argued that Canada displayed international leadership in Montreal and "must not shift from leaders to global villains without a vigorous public debate."[46] Another environmentalist expressed concern that the Conservative opposition to climate research suggested Harper would move Canada "into the same camp as U.S. President George W. Bush."[47] This fear was not only voiced by Canadian environmentalists. A BBC News editorial by Richard Black following the January 2006 minority election of the Conservatives proposed that in a "history of the Kyoto Protocol" Canada will be seen as particularly influential.[48] Not only did Canada's Liberals nurse "the treaty back from the brink of oblivion" by ratifying it in 2002 and then pressure for the 2005 Montreal Action Plan but the new prime minister's statements that Kyoto is "a terrible treaty" and a "wasteful exercise" implied Canada would now shift the international response toward the position of its southern neighbour.

RESISTING PROGRESS

The political actions of April 2006 did not instil optimism in those who were concerned newly elected Prime Minister Harper would align Canada with America's denial of climate change. His government's first act stopped funding to an education program that aimed to help people cut greenhouse gas emissions, as well as place under review "100 other climate change programs."[49] On May 2, 2006, the government's first budget replaced the Liberal plan with expensive tax credits for transit passes, while not mentioning the Kyoto Protocol, climate research, or the need to promote renewable energy initiatives.[50] The preference for a more market-friendly approach was further highlighted in October with the tabling of the national Clean Air Act. As was mentioned in the previous chapter, it did not mention the Kyoto Protocol and proposed using intensity targets and technology adaptations to reduce emissions by 31 to 56 percent of 2003 levels by 2050. If this new policy did not reveal the government's primary concern, the prime minister's public engagement on the day of the Clean Air Act's release made it abundantly clear. At a convention of insurance brokers, Harper stated that when speaking to international investors "the most important sector story I have to tell is energy" for "Canada is an emerging energy superpower."[51] Looking at a world where demand for resources is increasing as India and China rush toward industrialization, he concluded the potential that Alberta's tar sands offers "Canada's long-term economic growth is truly breathtaking."[52] The denial of the Friends of Science and the American Republican government seemed to be making the failed Liberal call for a global conscience a distant Canadian memory.

This climate policy change was part of Prime Minister Harper's overall agenda of revolutionizing the nation's dominant Liberal approach to government. His Conservative Party was created in 2003 by bringing together the preceding Progressive Conservative party's national focus with the Alliance/Reform party's goal of achieving political leverage for Western Canadian provinces like Alberta that was "equivalent to the region's contribution to the national economy."[53] Until their defeat in 2006, this approach was dismissed by Liberals as not only regional politics with little national applicability but also as an American-style neo-conservativism that contradicted Canadian values. In the words of political economist Brian Cooper, Liberal Canada disregarded Albertan perspectives by characterizing them as those belonging to unsophisticated right-wing cowboys from Texas North "who care little about their fellow Canadians."[54] In contrast to such a dismissive view, he characterizes Prime Minister Harper's modest dissent as a revolution that aims to correct Canada's progressive Liberal history of attempting "to improve economic and social well-being through higher spending" and applying national standards to policy areas that belong to the provinces.[55] Cooper's insights are relevant not only because it was out of his University of Calgary political economy department that Prime Minister Harper graduated, but also because his thought resonates with that of the Friends of Science. Considering the previous chapter's analysis, it should not be surprising that his book argues Environment Canada had, during the Liberal regime, funded "research on so-called anthropogenic climate change because it is an article of faith" rather than a scientifically verifiable phenomena.[56]

One central conclusion of Cooper's analysis was that the Harper government would meet significant blocks in enacting its 2006 revolution because of a liberal "administrative tyranny" that had built up in the bureaucracy after years of Liberal government.[57] The month prior to the Conservative unveiling of its Clean Air Act, this bureaucratic liberal resistance presented itself in an audit of Canada's climate response that was submitted to the House of Commons by the national Commissioner of the Environment and Sustainable Development, Johanne Gélinas.[58] If the Conservative government is correct in proposing that "Canada cannot realistically meet its Kyoto target," then Gélinas proposed it is time to develop "new targets" that attempt to at least "slow the rate of growth of greenhouse gas emissions."[59] The report further challenged the prime minister to state how his government will reduce greenhouse gas emissions in a context that has seen oil and gas sector emissions increase "over 50 percent since 1990" and those of the Alberta tar sands, which are projected to "double between 2004 and 2015."[60]

While Cooper and the Friends of Science would see such an analysis as reflective of liberalized bureaucracy and science, the commissioner was equally critical of the Liberals who had governed when much of Canada's climate policy was enacted. Partially supporting the Friends of Science view, Gélinas reported that Liberal policies were highly ineffective for a number of reasons. First, after spending over $6 billion, the bureaucracy still did not "have an effective government-wide system to track expenditures, performance, and results on its climate change programs."[61] Second, it was highly unlikely that following through on the Liberal emission reduction policies would have met their Kyoto goals, with the best case scenario being they might "slow the rate of growth."[62] In contrast to the position held by Cooper and the Friends of Science, which suggested human-based climate change was unproven, the commissioner offered a more scientifically nuanced sense of those economic predispositions that were limiting both Conservative and Liberal approaches to a valid climate change issue.

One way of thinking about Canada's Conservative shift from the Liberal position on climate change is to consider the political economic thought of Karl Polanyi. In a historical analysis of the West's growth economy, he identified a political "double movement" that is defined by the "action of two organizing principles in society, each of them setting itself specific institutional aims, having the support of definite social forces and using its own distinctive methods."[63] Looking at England's social changes during the Industrial Revolution, Polanyi describes a process by which the "formal" market system extended its influence by dispossessing people of their "substantive" approaches to "redistribution, customs of reciprocity, householding, forums of non-profit exchange."[64] With this progressive displacement of substantive ways over the nineteenth and twentieth centuries came the transformation of people into labour and nature into resources and services that could be economically managed. The result was an industrially growing economy, but one with depressive conditions that created class differences analogous with today's social and environmental realities of developed and developing nations. On one side of this double movement was the economic liberal attempt to establish a self-regulatory market that depended on the trading classes and free trade, while on the other was a conservative cultural movement that responded with the aim of protecting the pre-existing human and natural bonds. The difference in the contemporary Canadian dynamic is that Liberals are concerned with sustaining formal economic progress while mitigating the degree of climate impacts, and Conservatives want to conserve the pre-existing cultural bonds that have been built upon the once progressive tradition of industrial economic growth.

While the economizing tendency at the core of the Conservative revolution is inconsistent with Polanyi's analysis, something closer to his substantive resistance can be sensed by shifting our perspective from Harper's Western movement to critiques arising from the Inuit North. The substantive ways referred to by Polanyi can be discerned in Arnakak's definition of IQ's four central premises.[65] The first is that "the extended family is the primary life support system," though this fundamental unit is embedded within broader human and ecological communities. Consequently, he explains that IQ's second defining feature attempts to ensure that economic actions are oriented toward the optimization of social and ecological networks as opposed to focusing upon an individual's economic self-interest. This is consistent with what the first chapter described as the contextualizing value of *Silatuniq*. The third premise asserts that "the family is the means of transferring knowledge and skills that make sense, and are sustainable in the context within which they arose." The final premise that Arnakak describes is that in IQ the extended family "is the basic economic unit of society," with its actions being based in an egalitarian sharing system that assumes people have the right to basic human needs. The underlying value of this substantive approach seems to be sharing rather than economizing, and it can be seen influencing the current Inuit relation with Canadian and international forces.

When the Inuit Circumpolar Conference confronted the *Arctic Climate Impact Assessment* for not going far enough in its political response to northern warming, it highlighted various research projects that it felt represented a historic shift away from the progressive marginalization of Inuit views. The Inuit Circumpolar Conference explains that in this new research, indigenous knowledge has been "accepted as legitimate, accurate, and useful, although until recently it was dismissed by some scientists as anecdotal and unreliable."[66] Thinking about a response to northern warming, the Inuit Circumpolar Conference added that "Inuit have repeatedly offered to share what they know of their environment in the expectation that their observations will assist governments to manage natural resources" in a way beneficial to both Western nations and Inuit.[67] This intercultural dimension of a substantive sharing approach also seemed to inform the 1999 creation of Canada's Inuit territory of Nunavut, which in Inuktitut means "our land." Examining the Inuit strategy for attaining this political control, historian J. R. Miller explains its leaders couched their proposals in low-key terms that disarmed critics by suggesting "Inuit simply wanted into the Canadian political family."[68] The genius of this approach is "that it enables southerners to see Inuit victory as a triumph that does not represent a loss for the rest of the country."[69] While the Conservative embrace of IQ at the Polar Bear Roundtable can be seen as another victory for this Inuit approach, it has occurred in a context of Conservative denials and Liberal failures to manifest a global conscience that is responsive to *Sila*'s northern warming or Gaia's climate changes.

In contrast to the substantive approach of Inuit, Western policy and research has generally promoted a formal economic sense of development that has historically undermined many indigenous cultures, their knowledge, and the ecologies to which they are related. Offering an IQ critique of formal economic growth, Arnakak describes it as a cultural machine that works "well with maximal settings because they're simple and linear," but are ultimately toxic "for living breathing forms."[70] Similarly, the Polanyi perspective of Ellen Wood concludes that because "the destructive effects of capitalism are outstripping its material gains," the continued expansion of this system will likely result in more dispossession, destruction, and receding material benefits.[71] Over the past couple decades, the liberal response to this ecologically and socially toxic situation has been to implement a "sustainable development" agenda that can stabilize environmental issues and sustain economic growth.

Such a progressive movement typifies what political economist Colin Hay describes as the "logic of crisis displacement," for rather than viewing these crises as a function of economic growth, the liberal approach characterizes them as "political crises within capitalism."[72] The central issue is echoed by Christopher Lasch, who argues that liberalism's greatest contradiction is its aim to sustain economic growth while promoting equity in a global context where resources "hitherto imagined to be inexhaustible, are already approaching their outer limit."[73] Returning to the IQ view of Arnakak, he proposes that making sustainable development feasible will first require "a complete reframing of the current formal economic system" so that the focus on a profitable bottom line is secondary to sustaining cultural "integration into its particular ecological context."[74] The difficulty is that such contextualizing runs counter to the Western belief in a formal economic system based on individual self-interest. As Lasch concludes, the progressive tradition has fostered a cultural belief that denies there are any "natural limits on human power and freedom."[75] In Canadian politics, this "formal" economic predisposition has not only become the focus of cultural conservation for Prime Minister Harper's revolution but it was also a core feature of the failed Liberal climate response.

At the 1997 Kyoto negotiations, the pervasiveness of this formal economizing approach in the deliberations led the World Council of Churches to release a statement that depicted climate change as a call from God to "challenge the injustice and contribute to alternative approaches."[76] While we have seen that Inuit perspectives on the changing *Sila* can likewise challenge Canada's and the West's economizing approach, the Gaia theology of Anne Primavesi suggests the blame for our political economic and cultural failure of conscience goes far beyond the difficulties of interdependent national economies.[77] Rather, she argues it is primarily due to the continued inability of people—including politicians, climate researchers, and theologians—to conceive themselves as part of Gaian processes. This is an issue that eco-theologian Stephen Scharper has likewise considered in relation to the limitations of Gaia theory itself. After considering Lovelock's dismissal of the ozone depletion problem in the 1980s, Scharper came to the conclusion that the theories of scientists can lead them to "ignore data and minimize mammoth problems which belie their visions."[78]

Beyond highlighting the cultural limits to scientific objectivity, Scharper's deeper concern is that scientific conceptions of Gaia display a severely limited appreciation of social justice issues—a point that is also made by Primavesi. As Scharper states, there is next to no critique of "existing power structures as well as historical patterns of inequality" that are the context of today's political irresponsibility to issues like climate change.[79] Such an analysis does not mean Gaian indeterminism cannot include a consideration of injustices, only that an interdisciplinary project focused predominantly on physical science and economic development has just begun to think through the way in which these social and environmental issues interact. With a human-centred and globally economizing mentality continuing to block our conception of today's climate changes, Scharper, Primavesi, the World Council of Churches and Arnakak can be seen as challenging all of us to enact a "change of mind" that will have practical economic consequences for human relations with Gaia and its many regional ecologies.

Though such a spirited change in our economic mindset is probably not what Prime Minister Martin had in mind when he told the Montreal delegates, "We are in this together" and that the "developed world cannot walk away from its responsibilities,"[80] it does seem as though the West's formal economic

influence on northern warming will eventually require nations like Canada to take responsibility for a planetary change
of mind. The alternative is to be complicit in fuelling climate
changes that anthropologists Susan Crate and Mark Nuttall characterize as "environmental colonialism at its fullest
development."[81] As they explain, the global greenhouse gas
processes have not been initiated by people who live in the cli-
mate-sensitive regions of the world, such as the North, yet these
warming changes will magnify "already existing problems of
poverty, deterritoriality, marginalization, and noninclusion in
national and international policy-making processes and discourses."[82] More than that, there will be a continued degradation of *Silatuniq* traditions because of the way in which they are
related to a quickly changing ecology. Unless the marginalization of substantive ways becomes a central concern of climate
research and policy, it is highly likely Inuit, as well as many
other marginalized cultures, will continue to be undermined
and Canada will continue to fail at enacting a global or northern conscience.

CONCLUSION: A COLONIAL SYMBOL

When Prime Minister Martin held Montreal up as a potential
international symbol for environmental progress, he clearly
did not consider the city's historic involvement in colonial
dynamics as a potential factor that is limiting Canada's enactment of a global conscience. The recent Conservative concern
with a melting Northwest Passage is in some ways rooted in
Montreal's past, for it was here that Jacques Cartier's second
1535 search for a route to the east came to its conclusion before
being re-initiated with Samuel de Champlain's 1613 survey of
Montreal Island. It was only during the latter half of the seventeenth century that Montreal became a village of fur traders
more concerned with the wealth in the surrounding land
than the wealth beyond Canada. Being a hub for the fur trade,

Montreal came to historically symbolize Canada's still predominant "frontier economy" approach to the environment as a supplier of resources.[83] This approach has changed over time as the pre-1800 focus on securing property and resource rights from indigenous groups gave way to the nineteenth-century concern with licensing companies to rapidly extract resources. As the twentieth century approached, governments began moderating the rate of resource extraction, at first because of long-term considerations, and then as a management response to environmental concerns that began presenting themselves in the 1950s and 1960s. Throughout this changing history, Canada has maintained a primary concern with extracting resources in a vast land rather than thinking about "the destruction of the environment."[84]

In many ways, the Conservative revolution that is concerned with the economic prospects of the Northwest Passage and the tar sands is grounded in conserving this Canadian frontier culture that for a time radiated from Montreal. As for the sustainable development agenda promoted by Liberals at the Montreal conference, there is one significant commonality with the frontier economic tradition: they both utilize specific types of science that are consistent with formal economic plans while sidelining substantive ethical and cultural views.[85] The difficulties of such formal frontier economics go far beyond the Canadian relation with the surrounding environment, for Arnakak writes current perspectives on IQ reflect "a thinly veiled corporatist agenda regarding environmental and resource development" that are far too narrow "to really be considered indigenous knowledge."[86] There is something about the West's current climate research and policy tendencies that undermines the engagement of other types of cultural knowledge like IQ. Considering these limiting cultural assumptions, Fikret Berkes argues Western research is dominated by a liberal Enlightenment "belief that all cultures would merge into one 'correct' way of thinking about human development and well-being."[87]

Social theorist Bruno Latour further connects this Western tendency to two divisions that were inaugurated during the Enlightenment.[88] The first great historic divide allowed science to separate nature from culture so as to experimentally manipulate natural laws and economically calculate progress. From this objective positioning of human over nature, the West instituted a second great intercultural divide that marginalized other types of cultural knowledge because of their perceived incapacity to objectively separate "what is knowledge from what is Society."[89] Despite the many negative social and ecological impacts related to succeeding colonial and industrial agendas, Latour states Western culture still views itself as escaping "from the prison of the social or of language to gain access to things themselves through a providential exit gate, that of scientific knowledge."[90] The past couple decades have seen this approach amended to the extent that indigenous traditional ecological knowledge is now acceptable, but, as Arnakak, Berkes, and others suggest, such a narrow approach merely continues the rationalization of substantive ways of living. What is particularly valuable in Latour's analysis is that he clarifies that the difficulties facing Western climate research are related to its marginalization of both other types of cultural knowledge and its own ethical and religious traditions. In contrast to the reduced historic perspective of Prime Minister Martin's progressive Montreal, these analyses suggest the colonial past of this city offers a historic symbol of those formal economizing trends that are integral to Canada's failed global and northern conscience.

Early in the Chesterfield Inlet workshop, Eli Kimmaliardjuk expressed a now familiar frustration at the lack of knowledge-sharing that has shaped the colonial relation between Western researchers and Inuit, to which Louis Autut added: "We really don't believe scientists anymore because they never report anything." He followed this by asking a couple of poignant questions: "Why don't they give information to us and why don't they want to know from us?" Building upon his questions,

Kimmaliardjuk suggested that after an elder "dies the scientists will take what he says and use it, but they will not give credit." Similar concerns were expressed in Sachs Harbour where Riedlinger and Berkes found Inuit are frustrated "over the lack of useful feedback from scientists or input into research."[91]

Such a Western approach that undermines a complementarity of cultural perspectives on climate change is not only a common current Inuit experience, but is part of a long colonial history that has been connected here to Inuit *ilira* and *Sila's* northern warming. The IPCC most definitely does not make such a direct connection between colonialism and climate change, but it does suggest the possibility of such an analysis when it states that "shifts or declines in resources resulting in altered access to subsistence species (e.g., Inuit hunting of polar bear) can lead to rapid and long-term cultural change and loss of traditions."[92] As was touched upon in relation to the historic narrowing of Inuit views on *Sila*, Western assumptions and politics have manifested colonial cultural changes that Crate and Nuttall suggest are historically connected to today's northern warming. In the next chapter, I historically extend these leads in relation to the Inuit experience by analyzing the connection between colonialism, formal economic logic, and climate change in the context of early-twentieth-century contact around Chesterfield Inlet and the Hudson Bay area.

ENDNOTES

1   Paul Martin, "Address by Prime Minister Paul Martin at the UN Conference on Climate Change," Montreal, QC, 7 December 2005, http://pm.gc.ca/eng/news.asp?id=666.

2   Ibid.

3   CBC News, "Martin Urges Nations to Get Tough on Energy Consumption," 7 December 2005, http://www.cbc.ca/story/canada/national/2005/12/07martinclimate.

4   Martin, "Address by Prime Minister Paul Martin at the UN Conference on Climate Change."

5   Steven Bernstein, "International Institutions and the Framing of Canada's Climate Change Policy: Mitigating or Masking the Integrity Gap?" in E. Lee and A. Perl, eds., *The Integrity Gap: Canada's Environmental Policy and Institutions* (Vancouver: UBC Press, 2003), 87.

6   Anthony Perl and Eugene Lee, "Conclusion," in E. Lee and A. Perl, eds., *The Integrity Gap: Canada's Environmental Policy and Institutions* (Vancouver: UBC Press, 2003), 253.

7   Eugene Lee and Anthony Perl, "Introduction: Institutions and the Integrity Gap in Canadian Environmental Policy," in E. Lee and A. Perl, eds., *The Integrity Gap: Canada's Environmental Policy and Institutions* (Vancouver: UBC Press, 2003), 23; Perl and Lee, "Conclusion," 241.

8   Bernstein, "International Institutions," 77.

9   Martin, "Address by Prime Minister Paul Martin at the UN Conference on Climate Change."

10   Mark Nuttall, "Indigenous Peoples' Organizations and Arctic Environmental Cooperation," in M. Nuttall and T. V. Callaghan, eds., *The Arctic: Environment, People, Policy* (Amsterdam: Harwood Academic Publishers, 2000), 621.

11   Habiba Gitay et al., "Ecosystems and Their Goods and Services," in contribution of Working Group II to the TAR of the IPCC, *Climate Change 2001: Impacts, Adaptation and Vulnerability* (Cambridge: Cambridge University Press, 2001), 277.

12 E-mail correspondence, 22 March 2004.
13 Martin, "Address by Prime Minister Paul Martin at the UN Conference on Climate Change."
14 Bernstein, "International Institutions," 87.
15 Ibid.
16 Ibid.
17 Quoted in Spencer Weart, *The Discovery of Global Warming* (Cambridge: Harvard University Press, 2003), 172.
18 Bernstein, "International Institutions"; also see Heike Schröder, *Negotiating the Kyoto Protocol: An Analysis of Negotiation Dynamics in International Negotiations* (Piscataway, NJ: Transaction Publishers, 2001).
19 Bernstein, "International Institutions," 90.
20 Heike Schröder, *Negotiating the Kyoto Protocol: An Analysis of Negotiation Dynamics in International Negotiations* (Piscataway, NJ: Transaction Publishers, 2001), 79; also see Weart, *The Discovery of Global Warming* and Lydia Dotto, *Storm Warning: Gambling with the Climate of Our Planet* (Toronto: Doubleday Canada, 2000).
21 Bernstein, "International Institutions," 90; Judith McKenzie, *Environmental Politics in Canada: Managing the Commons into the Twenty-First Century* (Don Mills: Oxford University Press, 2002).
22 Weart, *The Discovery of Global Warming*, 174; these are issues are also covered in Stephen Schneider and Jose Sarukhan, "Overview of Impacts, Adaptation, and Vulnerability to Climate Change," in *Climate Change 2001: Impacts, Adaptation, and Vulnerability*, contribution of Working Group II to the TAR of the IPCC (Cambridge: Cambridge University Press, 2001); Barry Smit and Olga Pilifosova, "Adaptation to Climate Change in the Context of Sustainable Development and Equity," in *Climate Change 2001: Impacts, Adaptation, and Vulnerability*, contribution of Working Group II to the TAR of the IPCC (Cambridge: Cambridge University Press, 2001).
23 Quoted in Gustavo Esteva, "Development," in W. Sachs, ed.,

The Development Dictionary: A Guide to Knowledge as Power (London: Zed Books, 1992), 6.

24 José M. Sbert, "Progress," in W. Sachs, ed., *The Development Dictionary: A Guide to Knowledge as Power* (London: Zed Books, 1992), 194.

25 Gilbert Rist, *The History of Development: From Western Origins to Global Faith* (London: Zed Books, 2002), 79.

26 Gro Harlem Brundtland, *Our Common Future* (Oxford: Oxford University Press, 1987), 8.

27 Rist, *The History of Development*, 186.

28 Weart, *The Discovery of Global Warming*, 174; Schneider and Sarukhan, "Overview of Impacts, Adaptation, and Vulnerability to Climate Change"; Smit and Pilifosova, "Adaptation to Climate Change in the Context of Sustainable Development and Equity."

29 David G. Hallman, "Ethics and Sustainable Development," in D. G. Hallman, ed., *Ecotheology: Voices from South and North* (Maryknoll, NY: Orbis Books, 1994), 269.

30 Weart, *The Discovery of Global Warming*, 174.

31 Quoted in McKenzie, *Environmental Politics in Canada*, 232.

32 Mark Jaccard et al., *The Cost of Climate Policy* (Vancouver: UBC Press, 2002), 198.

33 Perl and Lee, "Conclusion," 247.

34 McKenzie, *Environmental Politics in Canada*, 231.

35 Quoted in Bernstein, "International Institutions," 96.

36 Government of Canada, *Moving Forward on Climate Change: A Plan for Honouring Our Kyoto Commitment*, 13 April 2005, http://www.climatechange.gc.ca, 12.

37 Gary Doer and Jean Charest, "Seize the Climate-Friendly Day," *Globe and Mail*, 7 December 2005, sec. A, p. 27.

38 Government of Canada, *Moving Forward on Climate Change*, iii.

39 Donald Stanley Lemmen, Fiona J. Warren, and Climate Change Impacts and Adaptation Program, *Climate Change*

Impacts and Adaptation: A Canadian Perspective (Ottawa: Climate Change Impacts and Adaptation Program, 2004), 25.

40 Ibid., 25.

41 Richard S. Tol, "Is the Uncertainty about Climate Change Too Large for Expected Cost-Benefit Analysis?" *Climatic Change* 56 (2003): 282.

42 Smit and Pilifosova, "Adaptation to Climate Change in the Context of Sustainable Development and Equity."

43 Weart, *The Discovery of Global Warming*, 138; David Demeritt, "The Construction of Global Warming and the Politics of Science," *Annals of the Association of American Geographers* 91, no. 2 (2001): 307–337.

44 United Nations Framework Convention on Climate Change, *Key GHG Data*, 2005, http://unfccc.int.2860.php.

45 Ibid.

46 Sierra Club of Canada, "Harper's Position on Kyoto: A Tragedy for the Planet," 12 January 2006, http://www.sierraclub.ca.

47 Sierra Club of Canada, "Canadian Climate Coalition Denounces Conservative Party for Ducking the Issues," 17 January 2006, http://www.sierraclub.ca.

48 Richard Black, "Will Kyoto Die at Canadian Hands?" *BBC News*, 27 January 2006, http://www.bbc.co.uk.

49 Martin Mittelstaedt, "Ottawa Stops Funding One Tonne Challenge," *Globe and Mail*, 1 April 2006, sec. A, p. 7.

50 David Suzuki Foundation, "News Release: Federal Budget Fails to Mention Global Warming," 2 May 2006, http://www.davidsuzuki.org/climate_change.

51 Gloria Galloway, "PM Gives Strong Defence of Energy Sector," *Globe and Mail*, 20 October 2006, sec. A, p. 4.

52 Ibid.

53 Roger Gibbins and Sonia Arrison, *Western Visions: Perspectives on the West in Canada* (Peterborough, ON: Broadview Press, 1995), 41.

54   Barry Cooper, *It's the Regime, Stupid! A Report from the Cowboy West on Why Stephen Harper Matters* (Toronto: Key Porter Books, 2009), 125.

55   Ibid., 156–157, 187.

56   Ibid., 144–145.

57   Ibid., 227.

58   Commissioner of the Environment and Sustainable Development, *The Commissioner's Perspective 2006: Climate Change, An Overview and Main Points* (Ottawa: Office of the Auditor General of Canada, 2006).

59   Ibid., 13.

60   Ibid., 12.

61   Ibid., 10.

62   Ibid., 9.

63   Karl Polanyi, *The Great Transformation* (Boston: Beacon Press, 1957), 132.

64   Polanyi, *The Great Transformation*; also see Ellen Meiksins Wood, *The Origin of Capitalism* (New York: Monthly Review Press, 1999); Gregory Baum, *Karl Polanyi on Ethics and Economics* (Montreal: McGill-Queen's University Press, 1996).

65   E-mail correspondence, 26 February 2004; also see Jaypeetee Arnakak, *A Case for Inuit Qaujimanituqangit as a Philosophical Discourse* (Iqaluit, NU: JPT Consulting, 2004).

66   Sheila Watt-Cloutier, Terry Fenge, and Paul Crowley, *Responding to Global Climate Change: The Perspective of the Inuit Circumpolar Conference on the Arctic Climate Impact Assessment*, 16 February 2005, http://www.inuitcircumpolar.com.

67   Ibid.

68   J. R. Miller, *Skyscrapers Hide the Heavens: A History of Indian-White Relations in Canada* (Toronto: University of Toronto Press, 2000), 409.

69   Ibid., 410.

70   E-mail correspondence, 26 May 2004.

71   Wood, *The Origin of Capitalism*, 121.

72   Colin Hay, "Environmental Security and State Legitimacy," in M. O'Connor, ed., *Is Capitalism Sustainable? Political Economy and the Politics of Ecology* (New York: The Guilford Press, 1994), 219.

73   Christopher Lasch, *The True and Only Heaven: Progress and Its Critics* (New York: Norton, 1991), 529.

74   E-mail correspondence, 13 May 2004.

75   Lasch, *The True and Only Heaven*, 530.

76   David G. Hallman, "Climate Change: Ethics, Justice, and Sustainable Community," in D. T. Hessel and R. R. Ruether, eds., *Christianity and Ecology: Seeking Well-being of Earth and Humans* (Cambridge: Harvard University Press, 2000), 461, 468–469.

77   Anne Primavesi, *Gaia and Climate Change: A Theology of Gift Events* (London: Routledge/Taylor & Francis Group, 2009).

78   Stephen Scharper, "The Gaia Hypothesis: Implications for a Christian Political Theology of the Environment," *Cross Currents* 44, no. 2 (1994): 212.

79   Ibid., 218.

80   Martin, "Address by Prime Minister Paul Martin at the UN Conference on Climate Change."

81   Susan A. Crate and Mark Nuttall, "Introduction: Anthropology and Climate Change," in Susan A. Crate and Mark Nuttall, eds., *Anthropology and Climate Change: From Encounters to Actions* (Walnut Creek, CA: Left Coast Press, 2009), 11.

82   Ibid., 12.

83   M. Hessing, M. Howlett, and T. Summerville, *Canadian Natural Resource and Environmental Policy*, 2nd Edition (Vancouver: UBC Press, 2005).

84   Ibid., 55–56.

85   For this tendency in the frontier economic tradition, see ibid., 14.

86   E-mail correspondence, 8 July 2004.

87   Fikret Berkes, "Epilogue: Making Sense of Arctic Environ-

mental Change?" in I. Krupnik and D. Jolly, eds., *The Earth Is Faster Now: Indigenous Observations of Arctic Environmental Change* (Fairbanks, AK: Arcus, 2002); Fikret Berkes, *Sacred Ecology: Traditional Ecological Knowledge and Resource Management* (Philadelphia: Taylor & Francis, 1999), 178.

88   Bruno Latour, *We Have Never Been Modern* (Cambridge: Harvard University Press, 1993).

89   Ibid., 99.

90   Ibid.

91   D. Riedlinger and F. Berkes, "Contributions of Traditional Knowledge to Understanding Climate Change in the Canadian Arctic," *Polar Record* 37, no. 203 (2001): 315–328.

92   IPCC, *Climate Change 2007: Climate Change Impacts, Adaptation, and Vulnerability, Summary for Policymakers, Contribution of Working Group II to the Fourth Assessment Report of the IPCC* (Geneva, Switzerland: IPCC Secretariat/ World Meteorological Organization, 2007), 672.

COLONIAL APOLOGIES
FROM CANADA?

In a June 11, 2008, national address to the indigenous victims of Canada's residential schools, Prime Minister Stephen Harper apologized for the government's role and asked "forgiveness of the Aboriginal peoples of this country for failing them so profoundly."[1] After describing the Residential Schools Settlement Agreement and the Truth and Reconciliation Commission as the future foundation for indigenous-Canadian relations, he added that our new relationship will be based on "respect for each other and a desire to move forward together with a renewed understanding that strong families, strong communities and vibrant cultures and traditions will contribute to a stronger Canada."[2] Among the many indigenous leaders present for the apology was the Inuit representative Mary Simon, who responded in Inuktitut to demonstrate "that our language and culture is still strong." Continuing, she stated that the prime minister's apology filled her "with hope and compassion for my fellow aboriginal Canadians." While this hope for a new respectful relationship was in a way affirmed a few months later at the Polar Bear Roundtable by the government's embrace of Inuit Qaujimatuqangit (IQ), the connection of *Sila's* colonial narrowing to today's northern warming suggests the apology's intent may be too historically confined. As Julie Cruikshank's northern research finds, the melting ice, snow, and glaciers are releasing indigenous stories that equate the devastating impact of current environmental changes "with a history of colonialism and its imbalances."[3]

Though knowledge on the role of fossil fuels in climate change was in its infancy when eminent Canadian conservationist John Livingston wrote his 1981 book on northern oil development, his analysis is poignant because, as with Cruikshank, Crate, and Nuttall, it interconnects northern colonialism and industrial environmental issues to Western cultural forces.[4] Livingston explains that the Canadian history of religious, political, and formal economic missions hides their common grounding in a Western wasteland view of the North "as a land which has not as yet been enhanced by the hand of God through the tools of man."[5] The northern frontier had to be economically developed so that the land and its indigenous people could contribute to industrial progress. It was with the twentieth century's conversionary missions that the people and land began experiencing a succession of environmental issues related to both localized and distant industrial activities. Climate change is merely the latest northern environmental issue experienced by Inuit, as it was preceded by the biological effects of nuclear waste from the Cold War northern build-up and the bio-magnification of persistent organic pollutants in the meat-based diet.[6] In Chesterfield Inlet, I heard many associate climate change to not only *Sila*'s warming but also to pollutants, garbage, and, as will be discussed in Chapter 6, shifting animal behaviour. The colonial *ilira* of Inuit and the ethical critiques of Livingston and Cruikshank suggest such a view may be an accurate interpretation that situates today's warming in the colonial context of various co-emergent social and environmental issues.

Over the two days of discussion in Chesterfield Inlet, participants consistently related northern warming to various behaviours that were foreign to this land prior to its colonial relations. Many agreed with Andre Tautu's view that the North's present changes are related to the way "bags and oil containers are thrown away everywhere." Clarifying this concern, Bernie Putulik explained:

> Before I was born our culture said look after the land, don't pollute it, respect the animals, and you will get all the things you need. Our culture said not to leave the garbage on the

land as you will pollute it for the future, and as of right now it is polluted. Everywhere we go hunting, people leave their stuff. Before I was born, people did not leave barrels and stuff. This is what our culture told us before the Christians came.

Though the West's formal economic activities were seen as the most significant source of this pollution, statements like that of Putulik also suggest the legacy of Western colonial forces is impacting the sustainability of Inuit substantive ways of living. Such a view suggests Christian conversions were integral in making Inuit less knowledgeable of *Sila*'s multiple dimensions and less respectful of the land.

This Inuit assessment in a way seems consistent with the environmentalist critique that Inuit no longer embody their traditional culture and thus have no special claim for continuing the seal and, later, polar bear hunts. As will be seen, there are significant cultural differences underlying such assessments. Beginning with an analysis of the colonial experience in Chesterfield Inlet, this chapter will consider the validity of Cruikshank's conclusion that colonial contact is "not merely a story from the past, but one whose consequences continue to cascade through twenty-first century debates, such as those framing environmentalism, biodiversity, and global warming."[7] Through this analysis and subsequent chapters, I will climatically extend Prime Minister Harper's colonial apology as a means to envision a global conscience that can both recontextualize the West's formal economy and display *Silatuniq*.

CHESTERFIELD INLET'S CONVERSION

Long before Montreal symbolized Liberal environmental progress or the colonial dream of a frontier economy, it was seen as a place to convert the land's indigenous peoples. Its Christian roots stretch all the way back to 1535 when Jacques

Cartier's second of three voyages made camp for winter at the Iroquois settlement of *Hochelaga* on present-day Montreal Island. By February 1535 the cold and scurvy had killed about twenty-five men, and the French crew responded by placing a Virgin Mary image on a tree, across the snow and ice, to which all approached in procession.[8] It was over a century later, in 1642, that an explicit Christian mission was initiated by a group of pioneers who sailed to the former site of *Hochelaga* and built the town of Ville-Marie because it was seen as an optimal place for converting "Amerindians in what was then the remote hinterland."[9]

It did not take long for Ville-Marie to be transformed from a holy city dedicated to the Virgin Mary to a fur-trading centre "notorious for disorder and drunkenness" that was renamed Montreal in 1705.[10] Within this early eighteenth-century frontier town, the Grey Sisters were created by Marguerite d'Youville and were formally recognized in 1753 as the first Catholic Order of Nuns to originate in Canada. Their grey attire was a reminder of their practical work in providing refuge to the poor, elderly, and transient indigenous people, as well as highlighting the need to take responsibility for past humiliations. The latter lesson was one Mother d'Youville knew well, for her religious calling was partially in response to her deceased husband who had built his career on illegally giving alcohol to indigenous groups to get cheaper prices on furs.[11] Her Montreal mission was taken to Chesterfield Inlet in 1931 by four Grey Sisters who came to serve "the poor, the sick, and the abandoned," with Pope Pius XI calling it "the most beautiful, the most difficult, the most meritorious" mission.[12] Its success was reflected in the quick conversion of numerous Inuit around Chesterfield Inlet, and the vocational choice of one girl by the name of Poovlaleraq to become, in 1951, the first Inuit nun.[13]

A distinct Canadian mission to Chesterfield Inlet actually began with the arrival of the Oblate Catholic Order in 1912 and the Grey Sisters in 1931. Their missions were part of an early-twentieth-century trend that saw Canadian officials arrive in the North and the Inuit population suffer an approximate one-third loss due to foreign Western diseases. Despite their theological differences, the Catholic, Moravian, and Anglican missionaries were commonly dedicated "to changing the beliefs, and therefore many of the customs of all the Athabaskan and Inuit."[14] The conversions in Chesterfield Inlet highlight the importance of Harper's apology, for in the mid-twentieth century it was home to a regional residential school with which the Grey Sisters were involved. Reflecting on this past, Inuit documentary filmmaker Martin Kreelak remembers the school as having "old-fashioned missionaries" who taught young people there was "no room for Inuit traditions."[15] He describes how in the 1950s his family and many others from the western Hudson Bay region, who had not settled in communities as of yet, were forced by the Canadian government to relocate to places like Chesterfield Inlet. The immediacy of this relocation was also related by Andre Tautu, who recalled being raised by his grandparents on the land and that he had never seen "a wooden building until 1950 when we moved to the community." It was in this context that Poovlaleraq became a Grey Nun. What makes her story particularly interesting is that she was the daughter of Joseph Okatsiak, a shaman who missionaries often characterized as practicing sorcery and who Inuit traditionally recognized as having access to *Silatuniq*.[16]

In July of the same year that Poovlaleraq committed to the Grey Nuns of Montreal and was baptized Sister Pélagie, she wrote a letter to Oblate Father Thibert concerning a problem with the Christian conversions in Chesterfield Inlet:

> Although they are baptized, many Eskimos do not understand their religion very well as yet. They are those I must aid through my prayers and works, and especially the children, so they may understand what is good for them, follow the commandments and, in brief, lead a devout Christian

life. Our Divine Saviour has inspired me to direct and guide them on the right path.[17]

Based on research with the Tununermiut Inuit of northern Baffin Island, John Matthiason similarly found that Inuit continued practicing shamanism and animism while also becoming Christians.[18] Referring to the fast pace of this cultural change, Jaypeetee Arnakak was of the view that Christian conceptions of a soul struggling "for survival in a harsh, unforgiving environment" offered Inuit an easier way of dealing with their northern reality by simply being baptized and acknowledging "Christ as the rightful savior."[19] In fact, Matthiason explains the devotion traditionally given to the shamanistic religion was quickly converted to Christianity because missionaries were themselves "seen as powerful shamans who had personal relationships with the most powerful spirit helper of them all—Jesus Christ."[20] Writing from Pelly Bay in 1956, one Father recognized not only the Inuit tendency of seeing the priest as occupying the place "formerly held by the sorcerer," but also the way in which priests and sorcerers provided "a similar link between the spirits on one side and the Eskimos on the other."[21] In light of these conversion difficulties and the research that indicates many indigenous cultures viewed Christian missionaries as sorcerers,[22] it may be possible that Sister Pélagie's vocational choice was following the spirit of her father's profession. By becoming a Grey Nun, she could commune with the Christian God rather than *Sila* for inspired guidance on northern living during great cultural changes.

The different spiritual approaches of Inuit and missionaries were not totally foreign to each other. For instance, the Grey Sisters described a great Inuit ingenuity for making use of all that is available in the wilderness, which they saw as resonant with the teachings of Mother d'Youville on a "supernatural spirit of poverty" that makes good use of everything.[23] It was not until the mid-nineteenth century that the Grey Sisters launched their missionary efforts beyond colonial Montreal and into the wilderness of the Red River Settlement of present-day Manitoba, and later "the most remote territories of the Arctic Circle."[24] Referring to the Sisters who followed this discipline of Christian poverty to Chesterfield Inlet, Sister Fitts writes that in enduring "long months of darkness, bitter cold, almost complete isolation, the special self-annihilation of conformity to Eskimo ways," they have received God's benediction.[25] Despite this common recognition that *Sila*'s "great loneliness" can bring spiritual insight, there was clearly much that differed between the practices of a Catholic missionary and an Inuit shaman.

The Grey Sisters' approach to the wilderness partook in a long Catholic tradition that stretches back to the seventeenth-century French Jesuit mission around the St. Lawrence River and Great Lakes. Connecting their conversion efforts with Biblical stories of Cain's wilderness and Abel's civilization, the Jesuits described the land as a "barren, abandoned, and frequently hostile" wilderness and its indigenous inhabitants as uncultivated *sauvages*.[26] Such a distrust of nature and its related knowledge was dealt with through ascetic practices that were distinct from the Inuit shamanic tradition. These contrasting approaches were highlighted by Sister Pélagie when she reflected on her life as a Grey Nun:

> They have rules to follow, rules that are really hard, and Inuit don't have that kind of rules. We had to carry sharp objects that hurt the body, especially around the arms and the neck. I don't know how many nuns died trying to beat temptation.[27]

These methods for beating temptation and bringing union with the divine were as foreign to Sister Pélagie's shamanic heritage as Canada's colonial methods and the *ilira* they evoked. This was a different kind of spiritual practice or sorcery than the *Silatuniq* her father would have learned from the North's "great loneliness." It is interesting to note the Jesuit wilderness missions were accused of sorcery by many indigenous because of a confluence of signs that, in the twentieth century, would similarly impact the Inuit—wide-ranging illnesses, conversions, cultural erosion, and appropriation of the land's wealth.

For Inuit in mid-twentieth-century Chesterfield Inlet and elsewhere, these colonial changes were epitomized by the residential schools. Supporting Kreelak's experience, we saw in the first chapter that the Royal Commission on Aboriginal Peoples (RCAP) reported the children of residential schools were systematically resocialized in a way that aimed to sever them from the "artery of culture that ran between generations."[28] Through separation from parents and community, children were isolated in a world "hostile to identity, traditional belief and language."[29] The methods of residential schools included punishment, food deprivation, head-shaving and public beatings. One Inuit related to the commission the experience of having "a heavy, stinging paste rubbed on my face, which they did to stop us from expressing our Eskimo custom of raising our eyebrows for 'yes' and wrinkling our noses for 'no.'"[30] While the report clarifies there were caring staff, the power imbalance and isolation of the schools from indigenous and nonindigenous communities made them "opportunistic sites of abuse."

The school in Chesterfield Inlet became one such place where sexual abuse occurred, resulting in criminal charges and a public apology by the Catholic Bishop in the late twentieth century.[31] Diagnosing Canada's institutional factors that led to this situation, the RCAP states that the reluctance of governments to challenge Christian churches that were careless in dealing with abuse by their members was central to "the pattern of mishandling abuse." Though no charges were directed at the Grey Nuns, Kreelak was concerned about the relation of these abuses to Sister Pélagie leaving the Grey Sisters after twenty years of service. In response to this query, Poovlaleraq said: "It is not going to help anybody [going over those things] that led me to resign from being a Grey Nun. It will only help myself, that's my thoughts."[32] While Kreelak respects her reasons for keeping quiet, he suspects the residential school abuses were central to why she left the Order.

For many Inuit, the results of these *ilira* experiences were high levels of stress-related problems like homicide, social violence, child abuse, addiction, a suicide rate "3.9 times the national average," and an increasing intergenerational gap in cultural knowledge.[33] Reviewing these colonial impacts on Inuit culture drew me back to memories of the two years I spent as a government social worker in the Innu community of Davis Inlet on Labrador's northeastern coast during the late 1990s. This community, known to the Innu as *Utshimassit*, "the place of the boss," came into Canada's national consciousness and my own awareness in January 1993 after six children attempted suicide by barricading themselves in an unheated shack during -40°C temperatures and sniffing gasoline.[34] As with the other impacts of cultural stress, the act of gas-sniffing is intimately linked to the colonial experience. This response first became prevalent in indigenous communities in the 1960s, with it being the drug of choice because the isolation of northern communities makes common urban drugs less accessible.[35] When this situation is coupled with the finding that gas-sniffing tends to rise during "periods of increased conflict or disruptions,"[36] then it becomes understandable why it is prevalent. While a 1975 study of Cree and Inuit youth found that 62 percent "had sniffed gasoline at least once in the last six months,"[37] more recent research in Canada, the United States, and Australia has found 50 to 60 percent of indigenous children have sniffed gas at least once, and 11 to 50 percent "were acute or current petrol sniffers."[38] This behaviour is one symptom of a legacy of colonial practices like residential schools that resulted in many Inuit becoming, in the words of the RCAP, "stranded between cultures, deviants from the norms of both."

One reason Inuit could associate their first Christian missionaries with sorcerers was because they knew the *ilira*-evoking signs of such spiritual power well. Inuit shamans did not at all times and in all encounters conduct themselves in a way reflective of a *Silatuniq* that uses power for good. They could also undertake actions that were seen as evil.[39] As Joanasie Qajaarjuaq remarks, a bad shaman or sorcerer could kill others out of hatred, revenge, and jealousy; he could, in a sense, use the power of others by "making it known that he has the whole family in his power and can crush any member he chooses."[40] In a historical analysis of sorcery in cultures as diverse as the Inuit and pre-Enlightenment Europe, Wolfgang Behringer explains the common belief is that a powerful person with anti-social tendencies is capable of using forces beyond the understanding of others "to induce illness and death."[41] Offering a similar sense from a French European case study, ethnographer Jeanne Favret-Saada defines the sorcerer as one whose "domain is never big enough" to utilize all their power, and thus their surplus is used to appropriate the vitality of others.[42] These analyses are consistent with Inuit who knew shamans could potentially inflict harm on a particular individual or family. Looking retrospectively, it also seems consistent with the relation between powerful Western missions and an Inuit culture that has suffered deep social impacts. In a sense, colonialism replaced shamans with Christian, and then political, missionaries who used various practices to manifest severe cultural changes in Inuit substantive ways of living; then, as with all sorcery, the missionaries reaped the benefits in terms of personal, cultural, or national power.

The sorcerous appropriation of power is not simply an issue of the colonial past, for the analyses of Cruikshank, Livingston, and some Inuit allow us to interconnect these changes with the current destabilizing impacts of environmental issues like climate change. While in Chesterfield Inlet there was general agreement with Andre Tautu's statement that "I don't think the Arctic contributes to the climate changes," there was also support for Putulik's comments on the relation between *Sila*'s warming and the way in which Inuit have polluted the land following the colonial arrival of Christians. His words seemed to reflect a traditional indigenous view of sin as disorders that signify a divine moral response to human indiscretions, like unawareness or explicit sorcery.[43] Inuit translate sin as *piunngittuliniq*, which according to Arnakak, means "the creation of the not well-formed." His e-mail continued:

> Most Inuit still believe in that implicitly, and I suspect that this is the case in Chesterfield Inlet. One becomes ill and elders seek out moral cause through discussions that non-Inuit are ignorant of. They've been told outright that this understanding is evil and against church teachings.[44]

In many indigenous traditions, sin is viewed as "a failure to live up to one's responsibilities to the community."[45] From such a perspective, leaving garbage on a changing land that provides so much sustenance is an external sign of failed communal responsibility. Whereas awareness of sin traditionally requires indigenous people "to submit themselves to the rhythms of nature" through ritual practices,[46] a very different view of sinful behaviour has informed what Livingston refers to as the West's wasteland theology. To deepen our sense of what is needed in an apologetic global conscience, I want to now shift our perspective toward the historic Christian roots of a contemporary formal economic sorcery that is having broad climatic impacts.

WASTELAND SCARCITY

Acknowledging the Liberal party's role in this colonial history by being the governing power "for more than seventy years of the last century," Paul Martin's replacement, Stéphane Dion, stated that Liberals share responsibility for the residential schools and asked the nation to "face the truth to ensure that we never have to apologize to another generation."[47] With climate

change having potential social justice impacts, Dion also offered an environmental alternative to Prime Minister Harper's
denial. From the moment he was elected Liberal leader in December 2006 at the same Montreal conference centre where, as
Environment Minister, he had presided over the 2005 Montreal
climate conference, Dion proposed to lead the Liberal Party
and Canada toward an alternative "Three Pillar Approach" that
weaves "together, better than any other country in the world,
economic prosperity, social justice and environmental sustainability."[48] Over the following months, Prime Minister Harper
responded to Dion's green liberalism by announcing a slightly
more stringent plan to reduce greenhouse gas emissions by 20
percent, though intensity targets would be maintained. As well,
Canada would reach its Kyoto Protocol commitment of a 6 percent reduction of greenhouse gases below 1990 levels in the year
2025, thirteen years after the international agreement.[49] These
Conservative policy shifts did not impress environmentalists or
Dion, who described them as a fake option that abdicates our
responsibility to deal with climate change.[50] In much the same
spirit as his predecessor, Dion suggested that an effective and
just response will have to more fully embrace climate research
and policy than the Conservative approach. As I discussed in
the previous chapter, such a global conscience would also have
to transcend the limiting liberal progressive tradition. To conceive what this may look like, let us return to Livingston's conception of a wasteland theology that underlies Canada's formal
economy and interconnects colonialism with climate change.

Though today's climate changes are most often associated
with a Western fossil fuel dependency that began with the Industrial Revolution, the 2003 *State of the World* essay by Gary
Gardner followed other critiques that associate both the limited
engagement of intercultural knowledge and today's ecological
crisis with the Enlightenment's division of science and religion,
or, in Inuit terms, IQ and *Silatuniq*.[51] In much the same spirit
as Latour's analysis in the previous chapter, this division not

only gave Western culture a powerful natural and political economic science but also severed from the present a sense of how religious beliefs influenced the Enlightenment and thus inform today's ecological crisis. It was just such an analysis that Lynn White, Jr., offered in his seminal 1967 paper, which suggested the ecologically destructive "marriage between science and technology" was grounded in the Christian call for humanity to exploit nature.[52] Following White's thought, Livingston proposed that the Christian placement of fallen humanity between God above and nature below promoted the view that there is "no conceivable reason for the existence of the blue planet apart from the needs of God and man."[53] As with White, he concluded that Christianity's extreme anthropocentric vision of humanity's cosmic position fostered both the objective aims of mechanistic science and today's economizing tendencies. More importantly, Livingston's analysis suggests this wasteland theology informs the colonial conversion of Inuit and the current extension of formal economics as the dominant response to *Sila*'s northern warming.

It should not be surprising that it is in the cross-cultural work of anthropologists, such as Cruikshank and Wenzel, that concerns about the historic influence of Western economic and religious assumptions has been most rigorously highlighted. In 1996, eminent anthropologist Marshall Sahlins epitomized such an analysis in his examination of a Western economizing mind that has "survived long enough to become the main protagonist of all the human sciences."[54] This redirection of the anthropological discipline toward his culture arose because of the concern that this belief "bedevils our understandings of other peoples." It seems to be an issue that originated in a mistake Sahlins himself made in his influential essay, "The Original Affluent Society."[55] This 1972 review of hunter-gatherer research theorized that the hunting way of life is affluent because of the beneficial ratio between hours of work and leisure. In a critique of this theory, Nurit Bird-David argues that Sahlins's attempt to refute the assumption that Western culture is the epitome of affluence had one serious flaw.[56] By discussing "hunter-gatherers' work in terms of practical reason and ecological constraints," his analysis actually depended upon a Western economic model focused on the individual optimization of rational behaviour. Despite positing an alternative interpretation, Sahlins's early work was undermined by an economizing belief. His subsequent anthropological examination of the West's "native culture structures" reflected an attempt to deal with this tendency by challenging social scientists to turn their gaze on their own culture and uncover historic Christian assumptions that limit an understanding of other cultures and, as Livingston, White, and Gardner suggest, human-ecology relations.

Sahlins's 1996 analysis dealt with these issues by focusing on the way in which Christianity historically influenced economic rationality and, consequently, social scientific research.[57] Much of his analysis is dedicated to the Christian beginnings of Enlightenment's economic thought, beginning with Bernard Mandeville's complaint "that it was difficult to distinguish the obstacles to human endeavours that were due to man's body from those that come from the condition of the planet 'since it has been curs'd.'"[58] In 1714 Mandeville's *The Fable of the Bees* proposed that actions deemed vices by medieval Christians would produce the greatest public benefit.[59] Selfishness, greed, and acquisitive behaviour would increase society's wealth, and thus increase the well-being of all people. The paradox was that while these actions would benefit society as a whole, they were based on private vices that in the Christian tradition were reflective of humanity's participation in a fallen world. With the thought of David Hume and Adam Smith, Mandeville's paradox was transformed by the proposal that any vice that brought collective benefits is in fact a virtue. Hume's elucidation of economic scarcity was influential to Smith's conception of an "invisible hand" that virtuously increases the collective wealth of nations through the self-interested behaviour of individuals.[60]

Though Smith only made passing references to the invisible hand, it became a powerful theory because the evolving colonial trade of the seventeenth and eighteenth centuries brought new wealth to Europe that promoted interpreting "the market's transcendence of political boundaries as evidence of a natural force."[61] While supporting the positions of Sahlins, White, and Livingston, religion and economy scholar Richard Nelson also clarifies that this predilection to globalizing formal economic growth actually arose out of a Christian view of the world as having "only one true religion."[62] From within this religious cosmos, the diverse economic theories of Adam Smith, Karl Marx, and John Maynard Keynes commonly promoted the belief that economic truths are universal. The result has been that Western economists and politicians still assume "the correct economic answer can and should be applied to all kinds of circumstances and societies."[63] Even the secular progressivism of Liberals like Dion and Martin is, according to Nelson, grounded in a globalizing Christian postmillennialism that envisions a rationally manageable end—one that "must apply to all people in all places."[64] Exemplifying this assessment is Keynes's view that global progress would bring an "end of scarcity" within a few generations, thus allowing the self-interested behaviour underlying the formal economy to be transformed into something more befitting human potential.[65] Right to the present, economic rationality has, in Sahlins' words, transformed human misery "into the positive science of how we make the best of our eternal insufficiencies, the most possible satisfaction from means always less than our wants."[66]

It is not until the calculus of nineteenth-century neoclassical utilitarianism that economics begins to shake its association with Christian beliefs and becomes primarily defined "as the branch of social science that deals with the allocation of scarce resources among competing ends."[67] This rational invisible hand assumes there is a fundamental discord between the nearly infinite desires of individual consumers and producers, and the scarcity of means to fulfill that yearning. As Gustavo Esteva explains, the "law of scarcity was construed by economists to denote the technical assumption that man's wants are great, not to say infinite, whereas his means are limited though improvable."[68] The political economist Nicholas Xenos adds that scarcity is not merely a theory but a way of organizing social relationships that "resonates with our quotidian experience."[69] It is about the cost of consuming one thing and denying another, maximizing technical efficiency so satisfaction and profit are increased, and efficiently costing social conflicts arising because of society's inability to produce at the same rate as human desires. Despite its progressive dissociation from religion, Lionel Robbins's classic quote succinctly highlights its Christian ancestry:

> We have been turned out of Paradise. We have neither eternal life nor unlimited means of gratification. Everywhere we turn, if we choose one thing we must relinquish others which, in different circumstances, we would wish not to have relinquished. Scarcity of means to satisfy ends of varying importance is an almost ubiquitous condition of human behaviour.[70]

Being more explicit about the Christian roots of this economizing logic, Nelson proposes that original sin has today been transformed into natural dysfunctions that are "brought into existence by material scarcity."[71] He goes on to add that the necessary economic response has been to try and create a sinless world that is without scarcity in its "complete material abundance."[72] Such a paradise is clearly impossible because of Gaia's finite nature, not to mention the fact that certain regions like the northern wilderness are seen by the Western mind as materially impoverished—at least until northern warming results in the projected economic benefits of a melting Northwest Passage or increased access to fossil fuels.

Contemporary environmental issues have not fundamentally challenged this "native culture structure" of the West, as can
be seen in the formal economic conception of environmental
externalities. This concept refers to situations "when consumption or production of one economic unit enters into the utility
or production function of another economic unit without any
compensation,"[73] such as when Inuit ways are impacted by distant greenhouse gas emissions. A response to these unaccount-
ed costs was first considered in the early part of the twentieth
century by Arthur Pigou. His approach suggested governments
should take a liberal regulatory role in internalizing the negative externality by making the offending party liable for their
actions through taxation.[74] In the 1960s, Ronald Coase argued
against solving the problem through a political restraining of
the offending parties, and rather suggested making a decision
based on "whether the gain from preventing the harm is greater
than the loss which would be suffered elsewhere as a result of
stopping the action."[75] More recently, ecological economists
have argued for a response based on situating the economy
within ecological realities. In the popular *Natural Capitalism* of Paul Hawken, Hunter Lovins, and Amory Lovins, the
final reference is still left to an economic approach that places
everything on the balance sheet so that nothing is "externalized
because social or biological values don't 'fit' into accepted ac-
counting procedures."[76] Critiquing an earlier work by Hawken
that similarly concludes "every grain of sand will have to be
treasured" and counted, the student of Livingston and political
ecologist Raymond Rogers suggests this response amounts to
the market being "both the cause of and the solution to environmental problems."[77] Rather than *Sila*'s northern warming
and Gaia's climate changes calling into question the economic
logic that has brought forth today's cultural challenges, the concept of an externality re-entrenches a wasteland theology.

Though Prime Minister Harper's apology and alliance with
IQ at the Polar Bear Roundtable seems to reflect a new Can-

adian approach to Inuit and the North, his denial of climate research and enactment of the Clean Air Act partake in a kind of
Coasian externalizing approach that was further highlighted at
the 2007 Bali Climate Change Conference. This United Nations
gathering saw the Conservative government continue to bolster
American opposition to emission reductions through a number of actions. First, the government banished "environmentalists and opposition MPs from Canada's delegation"—despite
this being a long-lasting Liberal tradition—while also allowing
"an oil company and several business executives to join the official delegation."[78] During the negotiating session on the eve
of the agreement, then Environment Minister John Baird sent
a bureaucrat to replace him, while he attended a meeting with
other industrialized countries, like the United States, that were
working on an alternative agreement more in line with their
Clean Air Act.[79] Despite the European Union's hard stance on a
mitigation response, it was their agreement to drop "a target of
cutting emissions by 25 to 40 per cent in the world's wealthiest
nations by 2020" that was integral to securing the Bali Road-
map.[80] North America's conservative movement had further
delayed hard emission reduction targets by negotiating a further two-year period of talks but were forced to declare that
"deep cuts will be required in global emissions of greenhouse
gases to respond to the 'urgency' of the global warming crisis."[81]
These actions suggest Prime Minister Harper's colonial apology
to Inuit is historically limited by an approach to climate change
that projects the economic benefits of the tar sands and melting
Northwest Passage will outstrip the externalized costs of *Sila*'s
northern warming.

The need for a response to this externalizing economic
belief—or wasteland theology—is highlighted not only by
Canada's limited Conservative approach to climate change but
also by the way in which liberalized climate research has interpreted the potential impacts of northern warming on Inuit.
According to the Intergovernmental Panel on Climate Change

(IPCC), the costs of these changes may be minor for two reasons: "the populations of humans and other biota within polar regions are low," and these changes may "increase the overall productivity of natural systems."[82] The *Arctic Climate Impact Assessment* concurs that there may be economic benefits to northern warming, such as melting sea ice that will make real Prime Minister Harper's dream of a Northwest Passage, not to mention the projected agricultural benefits related to longer growing seasons and easier access to oil reserves.[83] Where this formal economic logic leads is not that surprising in light of the West's colonial legacy. Both the *Arctic Climate Impact Assessment* and the IPCC conclude that indigenous people who follow traditional substantive ways of living will have limited future adaptation opportunities, while more industrially developed and urbanized communities will adapt more readily as they are given access to new economic opportunities.[84] The implication is that the North's IQ and *Silatuniq* will continue to recede as Inuit are forced to adopt formal economic adaptations to climatic changes that are historically interconnected to the wasteland approach of Canada's colonial history.

To challenge this pervasive economizing faith, Rogers offers an environmental history that almost follows Sahlins's call for an anthropology of the West.[85] He looks at the European usury debates on the morality of charging interest that began in the medieval period and continued into the seventeenth century as a means for reconsidering the West's overly economic approach to environmental issues. Contrasting contemporary political, scientific, and environmental discourses, where economic concerns hold priority over ethics, Rogers demonstrates that at the beginning of the usury debate Christian ethical and religious interpretations superseded the emerging formal economic interests. The general concern at this time was, in Rogers's words, "that unchecked usury inverts the social order by making the world the Devil's so that people no longer know their place."[86] Confronting this rationalizing trend, which would be epitomized in Smith's invisible hand, these moral arguments about the sin of economic rationality suggested to Rogers a possible alternative response to its never-ending global growth. While sustainable development merely expands the global project through maintaining the primacy of economic rationality, the usury voices inspire a view of environmental issues as divine responses that call for an "externalizing of internalities" to contextualize self-interested economic behaviour. As with Inuit critiques of colonialism and Western research, these historic views suggest political economic tendencies need to be constrained by substantive ways that go beyond the economic urge to internalize. Perhaps Nelson was correct when he stated that the current era may "be defined by the theological answer given to the loss of faith in modern progressivism and its various offshoots."[87]

Recognizing the scientific, political economic, cultural, and religious uncertainties that inform environmental issues like climate change, Livingston concluded "there can be no technical answer to a moral problem."[88] In such an indeterminate situation it should not be surprising that, as with the Sahlins and IPCC examples, his broad environmental thought could likewise get caught in the universalizing wasteland of his Canadian and Western culture. While analyzing the environmentalist view on the Inuit seal hunt, George Wenzel found Livingston's position to be more nuanced but still problematic. He explains that according to Livingston the colonial introduction "of foreign artefacts and modes of exchange introduced 'a new way of life foreign to, and essentially antithetical to, aboriginal cultural traditions.'"[89] The result is a "neotraditional" cultural approach to northern living that exchanges fur for money, and in the process ensures that Inuit are forever situated at the extractive level of the formal economy. For Wenzel, this view has a very narrow sense of cultural adaptation that actually dismisses the validity of subsistence as a way of life and advocates the Inuit acceptance of assimilation into Canada's southern cultural predispositions.

It is a view consistent with the above IPCC statements on Inuit climate adaptation and the position of environmental scientists at the Polar Bear Roundtable. In the words of Wenzel, it presents Inuit as "willingly participating in the destruction of their own culture and the harmony of the ecosystem," and as such extends the colonial agenda in "denying to Inuit their past and a voice to their future."[90] This universalizing of Western scarcity, whether in Christian, political, environmental, or, more commonly today, economizing guises is the central feature of a wasteland theology. It is this Western belief that needs to be understood if we are to extend Harper's and Dion's colonial apology to the present issue of climate change, rather than leave it to a future apology for sorcerous damage already done—as with Canada's residential school survivors.

What I have tried to map out over the past two chapters is a scarce Western and, more specifically, Canadian view of Gaia and northern *Sila* that has influenced the Inuit experience of colonialism and climate change. It is interesting to note the contrast between this Western wasteland approach and the shamanistic tradition of Poovlaleraq's ancestry that entered the "great loneliness" to return with a contextualizing IQ and *Silatuniq*. This is not to say the two traditions are totally different, for they are both at least partially based on an individual experience that gives access to greater power. In Michele Stephen's anthropological review of sorcery, she interestingly associates the sorcerer's powerful individualized personality "to the pathology of the self in Western culture where selves are so rigorously bounded, or should be, that all must suffer from painful isolation."[91] Whereas the West's formal economy assumes the individual's self-interested desire to appropriate power is a human universal, she explains that for "cultures where selves are open and permeable to others, extraordinary measures are needed to create a will as individuated" as that of a sorcerer.[92]

Based on these intercultural resonances, Stephen redefines sorcery as a magical process for gaining power over others so as

not to "be engulfed by the flux that is material existence and social relationship."[93] While we have seen that Inuit indicate shamans in the past could use their power in such a sorcerous way,
the wasteland theology that informs Western and Canadian
formal economic responses to climate change can analogic-
ally be seen as fuelling a cultural sorcery. In other words, the
intent to promote unjust economic benefits for specific individuals and groups through externalizing climate impacts onto
others, like the Inuit, can be interpreted as a formal sorcery.
More than that, there is a magical quality to these injustices in
that the levers of power that industrially nurture Gaia's climate
changes are currently beyond the reach of Inuit, and the global
knowledge is for many—not all, as we have seen with the Inuit
Circumpolar Conference—beyond their IQ. To connect these
cultural dynamics with sorcery is to clarify that current formal
economic choices are not objective and ethically neutral but
reflect powerful cultural interests and predispositions. This is
what I think a climatic apology will need to address.

CONCLUSION: A FORGIVING CLIMATE

While Mother d'Youville founded the Grey Sisters over two
centuries ago as an order based on compassion for colonial
sufferers, the twenty-first-century theology of Anne Primavesi
asks us to extend such a conscience to the planet itself in this
time of spreading environmental injustices. Her call goes far
beyond suggesting humans should show compassion for Gaia's
many beings and cultures, for she also postulates today's climate
changes are initiating this change by moving the Western mind,
in almost regulatory fashion, toward "acts of forgiveness."[94] In a
recent book on the evolving Green Sister movement, Sarah Taylor describes contemporary nuns who are manifesting this Gaian call to eco-justice by blending their Catholic tradition with
what is known of today's deteriorating environment.[95] They talk
of many spiritual practices like coming to know the bioregional

spirit of where they live, but perhaps the most poignant in relation to this analysis is the call to "fast" from technological and
industrial goods as a means to detoxifying the body, land, and
planet. Such an approach is vastly different from the disciplined
torture of the body that Poovlaleraq found so different from her
Inuit traditions. As Taylor writes, "in the culture of green sisters there is a strong affirmation of the body" and the planet.[96]
Such a view is remarkably similar to Primavesi, who argues that
Gaian climate changes call for Western culture to fast from its
industrializing and economizing assumptions. Also consistent
with the Green Sisters is her view that the Christian wasteland
belief in a totally transcendent God needs to be tempered with
an immanent sense "of God as emerging from earthly knowledge."[97] What makes this a forgiving theology for Primavesi
is its grounding in the New Testament Greek view of forgiveness as "changing one's mind," with today's important change
being the greenhouse gas behaviours and related economizing
mindset that views the earth as being here "solely for our use
and benefit."[98]

Though there may indeed be a climate for a global conscience, it is clear from the preceding chapters that there are
still powerful forces limiting such a cultural change of today's
sorcery. A more animated sense of the limiting Western powers can be seen if we return to those indigenous survivors of
colonialism who turned to gas-sniffing. These youth describe
their drug choice as giving them "access to a different world"
than that of their culture's recent violent history,[99] but this alternative world is not anywhere close to the contextualizing
Silatuniq of a shaman like Poovlaleraq's father. With the fuel's
depression of the central nervous system comes aggressiveness,
long-term memory loss, and feelings of indifference that have
led Australian aborigines to describe gas-sniffers as those who
"cannot hear" because they do not comprehend the impacts of
their actions.[100] This is an apt way of describing the broader
colonial legacy in the North. Decades of being separated from

family, elders, and the land by residential schools left many Inuit physically and culturally disconnected from their substantive ways of living, ancestral IQ, and *Silatuniq*. Interestingly, Livingston similarly proposes that the industrial energy and economic infrastructure of Western society fosters pervasive "institutionalized delusions" that afflict environmental politics and research.[101] It is a diagnosis that suggests the cultural and political economic institutions arising from wasteland production have progressively restructured and narrowed the West's climate thought and responses. Perhaps it can even be proposed that many of us are "without ears" for considering the relation of Gaia's external climate changes to an internal cultural mindset because of the West's historic proximity to powerful fossil fuels. In this view, the ineffectiveness of Canada's Conservative denial and Liberal failure may be seen as grounded in the debilitating mental impact of emissions from the tar sands.

Before considering the proposed climate apology of Dion's Liberals in the final chapter, I will next expand my look at the way in which fossil fuel consumption in Canada and America is making it very difficult to hear a broad range of interdisciplinary and intercultural voices. We will follow those environmentalist critiques that have characterized Prime Minister Harper's climate and energy policies as "a recipe for delay" in the same mould as American President George W. Bush.[102] The importance of such a continental analysis goes beyond the superficial similarities between a former Republican President from oil-rich Texas and Canadian Conservatives grounded in Alberta, a province with the "second largest source of recoverable oil in the world, after Saudi Arabia."[103] More relevant is the fact that Alberta's tar sands are important to the global financial system because there is no instability related to war, as in the Middle East, and to the south is "the largest consumer of oil in the world, the United States."[104] Of historic significance is the way in which this flow of fuel partakes in a Canadian wasteland economy that saw its exports to the United States increase between 1940 and 2003 from 41 to 83 percent.[105] With American economic voices strongly influencing Canadian politics, research, and culture, it is important that we take a closer look at the now historic Republican defence of a North American fossil fuel tradition in the opening decade of the twenty-first century. This will allow us not only to further think through the formal economizing that today interconnects colonialism to northern warming, but will also offer a space to hear an American Christian view that can deepen our Western and Canadian sense of what an apologetic global conscience will entail.

ENDNOTES

1   Prime Minister Stephen Harper, "Prime Minister Harper Offers Full Apology on Behalf of Canadians for the Indian Residential Schools System," 11 June 2008, http://www.conservative.ca/EN/1091/100669.

2   Ibid.

3   Julie Cruikshank, *Do Glaciers Listen? Local Knowledge, Colonial Encounters, and Social Imagination* (Vancouver: UBC Press, 2005), 221, 250.

4   John A. Livingston, *Arctic Oil* (Toronto: CBC Merchandising, 1981).

5   Ibid., 115.

6   Sanjay Chaturvedi, "Arctic Geopolitics Then and Now," in Mark Nuttall and Terry V. Callaghan, eds., *The Arctic: Environment, People, Policy* (Amsterdam: Harwood Academic Publishers, 2000).

7   Cruikshank, *Do Glaciers Listen?* 148.

8   Joan Skogan, *Mary of Canada: The Virgin Mary in Canadian Culture, Spirituality, History, and Geography* (Banff, AB: Banff Centre Press, 2003), 63.

9   Peter N. Moogk, *La Nouvelle France: The Making of French Canada: A Cultural History* (East Lansing: Michigan State University Press, 2000), 255; W. J. Eccles, *The French in North America, 1500–1765* (East Lansing: Michigan State University Press, 1998).

10   Moogk, *La Nouvelle France*, 256.

11   Albertine Ferland-Angers and Grey Nuns, *Mother d'Youville: First Canadian Foundress: Marie-Marguerite Du Frost De Lajemmerais, Widow d'Youville, 1701–1771: Foundress of the Sisters of Charity of the General Hospital of Montreal, Grey Nuns* (Montreal: Sisters of Charity of Montreal "Grey Nuns," 2000), 136.

12   Quoted in Mary Pauline Fitts, *Hands to the Needy: Mother d'Youville, Apostle to the Poor* (Garden City, NY: Doubleday, 1950), 301.

13 Fitts, *Hands to the Needy*, 302–304.

14 Hugh Brody, *The Other Side of Eden: Hunters, Farmers and the Shaping of the World* (Vancouver: Douglas & McIntyre, 2000), 244.

15 Quoted in *Amarok's Song: Journey to Nunavut*, directed by Martin Kreelak and Ole Gjerstad, Words and Pictures Video, in association with the National Filmboard of Canada and the Inuit Broadcasting Corporation, 1998.

16 For Inuit examples, see Cornelius H. W. Remie and Jarich Oosten, "The Birth of a Catholic Inuit Community: The Transition to Christianity in Pelly Bay, Nunavut, 1935–1950," *Études/Inuit/Studies* 26, no. 1 (2002); John Stephen Matthiasson, *Living on the Land: Change among the Inuit of Baffin Island* (Peterborough, ON: Broadview Press, 1992).

17 Penny Petrone, *Northern Voices: Inuit Writing in English* (Toronto: University of Toronto Press, 1988), 133.

18 Matthiasson, *Living on the Land*, 24.

19 E-mail correspondence, 25 June 2004.

20 Matthiasson, *Living on the Land*, 120.

21 Quoted in Remie and Oosten, "The Birth of a Catholic Inuit Community," 112.

22 Remie and Oosten, "The Birth of a Catholic Inuit Community"; Matthiasson, *Living on the Land*.

23 Fitts, *Hands to the Needy*, 303.

24 Sister Marie Bonin, "The Grey Nuns and the Red River Settlement," *Manitoba History* 11 (Spring 1986), http://www.mhs.mb.ca/docs/mb_history/11/greynuns.shtml.

25 Sister Mary Murphy, "The Grey Nuns Travel West," *Manitoba Historical Society Transaction Series* 3 (December 1944), http://www.mhs.mb.ca/docs/transactions/3/ greynuns.shtml.

26 Carole Blackburn, *Harvest of Souls: The Jesuit Missions and Colonialism in North America, 1632–1650* (Montreal: McGill-Queen's University Press, 2000), 42; Also see Roger M. Carpenter, *The Renewed, the Destroyed, and the Remade: The Three Thought Worlds of the Huron and the Iroquois, 1609–1650* (East Lansing: Michigan State University Press, 2004).

27   Quoted in *Amarok's Song*.

28   RCAP, *Report of the Royal Commission on Aboriginal Peoples* (Ottawa: Indian and Northern Affairs Canada), http://www. ainc-inac.gc.ca/ch/rcap/sg/sgmm_e.html (accessed 10 November 2005).

29   Ibid.

30   Ibid.

31   Quoted in *Amarok's Song*.

32   Ibid.

33   Bernard Segal, "The Inhalant Dilemma: A Theoretical Perspective," in F. Beauvais and J. E. Trimble, *Sociocultural Perspectives on Volatile Solvent Use* (New York: The Haworth Press, Inc., 1997), 92; Myriam Denov and Kathryn Campbell, "Casualties of Aboriginal Displacement in Canada: Children at Risk Among the Innu of Labrador," *Refuge* 20, no. 2 (2002): 26; RCAP, *Report of the Royal Commission on Aboriginal Peoples*.

34   Denov and Campbell, "Casualties of Aboriginal Displacement in Canada," 27.

35   Sheree Cairney et al., "The Neurobehavioural Consequences of Petrol (Gasoline) Sniffing," *Neuroscience and Biobehavioral Reviews* 26 (2002).

36   Ibid., 82.

37   Cited in Denov and Campbell, "Casualties of Aboriginal Displacement in Canada," 26.

38   Cairney et al., "The Neurobehavioural Consequences of Petrol (Gasoline) Sniffing," 82.

39   *Diet of Souls*, directed by John Houston, Triad Films, Halifax, NS, 2004.

40   Quoted in John Bennett and Susan Diana Mary Rowley, *Uqalurait: An Oral History of Nunavut* (Montreal: McGill-Queen's University Press, 2004), 194.

41   Wolfgang Behringer, *Witches and Witch-Hunts: A Global History* (Cambridge, UK: Polity Press, 2004), 13.

42   Jeanne Favret-Saada, *Deadly Words: Witchcraft in the Bocage* (Cambridge: Cambridge University Press, 1980).

43　Clara Sue Kidwell, Homer Noley, and George E. Tinker, *A Native American Theology* (Maryknoll, NY: Orbis Books, 2001); John W. Friesen, *Aboriginal Spirituality and Biblical Theology: Closer than You Think* (Calgary: Detselig, 2000).

44　E-mail correspondence, 2 November 2004.

45　Kidwell, Noley, and Tinker, *A Native American Theology*, 150.

46　Friesen, *Aboriginal Spirituality and Biblical Theology*, 111.

47　Stéphane Dion, "Residential Schools Apology," 11 June 2008, http://www.liberal.ca/story_14080_e.aspx.

48　Stéphane Dion, "Canada Will Not Fail the World: An Address by the Honourable Stéphane Dion," (speech delivered at the Liberal Leadership Convention, Montreal, QC, 1 December 2006).

49　Campbell Clark and Brian Laghi, "PM Charts a Greener Course," *Globe and Mail*, 5 January 2007, sec. A., p. 1.

50　Dion, "Canada Will Not Fail the World."

51　Gary Gardner, "Engaging Religion in the Quest for a Sustainable World," in L. Starke, ed., *State of the World 2003* (New York: W. W. Norton and Company, 2003), 153.

52　Lynn White, Jr., "The Historical Roots of Our Ecologic Crisis," *Science* 155, no. 3767 (1967): 1203, 1205.

53　John A. Livingston, *The John A. Livingston Reader: The Fallacy of Wildlife Conservation and One Cosmic Instant* (Toronto: McClelland and Stewart, 2007), 305.

54　Marshall Sahlins, "The Sadness of Sweetness: The Native Anthropology of Western Cosmology," *Current Anthropology* 37, no. 3 (1996): 395, 397.

55　For a critique, see Nurit Bird-David, "Beyond the Original Affluent Society: A Culturalist Reformulation," *Current Anthropology* 33, no. 1 (1992); for the original essay, see "The Original Affluent Society," in Marshall Sahlins's *Stone Age Economics* (London: Tavistock, 1972).

56　Bird-David, "Beyond the Original Affluent Society," 27.

57　Sahlins, "The Sadness of Sweetness."

58 Ibid., 396.

59 Bernard Mandeville, *The Fable of the Bees, Or, Private Vices, Publick Benefits* (Indianapolis: Liberty Classics, 1988).

60 Emma Rothschild, *Economics Sentiments: Adam Smith, Condorcet, and the Enlightenment* (Cambridge: Harvard University Press, 2001); Peter Minowitz, *Profits, Priests, and Princes: Adam Smith's Emancipation of Economics from Politics and Religion* (Stanford: Stanford University Press, 1993); Nicolas Xenos, *Scarcity and Modernity* (London: Routledge, 1989).

61 Joyce Oldham Appleby, *Economic Thought and Ideology in Seventeenth Century England* (Princeton: Princeton University Press, 1978), 279.

62 Robert H. Nelson, *Reaching for Heaven on Earth: The Theological Meaning of Economics* (Savage, MD: Rowman & Littlefield Publishers, 1991), 319; also see Robert H. Nelson, *Economics as Religion: From Samuelson to Chicago and Beyond* (University Park: Pennsylvania State University Press, 2001).

63 Nelson, *Reaching for Heaven on Earth*, 319.

64 Ibid., 7.

65 Ibid., 2.

66 Sahlins, "The Sadness of Sweetness," 397.

67 Ahmed M. Hussen, *Principles of Environmental Economics: Economics, Ecology and Public Policy* (London: Routledge, 1999), 6.

68 Gustavo Esteva, "Development," in W. Sachs, ed., *The Development Dictionary: A Guide to Knowledge as Power* (London: Zed Books, 1992), 19.

69 Xenos, *Scarcity and Modernity*, 2.

70 Lionel Robbins, *An Essay on the Nature & Significance of Economic Science* (London: Macmillan, 1952), 15.

71 Nelson, *Economics as Religion*, 28.

72 Ibid.

73 U. Sankar, *Environmental Economics* (Oxford: Oxford University Press, 2002), 3.

74 Ibid., 3.

75   Ibid., 4.

76   Paul L. Hawken, Hunter Lovins, and Amory B. Lovins, *Natural Capitalism: Creating the Next Industrial Revolution* (Boston: Little, Brown and Co., 1999), 319.

77   R. Rogers et al., "The Why of the 'Hau': Scarcity, Gifts, and Environmentalism," *Ecological Economics* 51 (2004): 181.

78   Geoffrey York, "Business Gets a Voice on Canadian Delegation," *Globe and Mail*, 10 December 2007, sec. A, p. 17.

79   Geoffrey York, "Baird a No-Show at Key Negotiating Session," *Globe and Mail*, 15 December 2007, sec. A, p. 2.

80   Geoffrey York, "Draft Bali Deal Omits Specific Targets," *Globe and Mail*, 15 December 2007, sec. A, p. 1.

81   Ibid.

82   O. Anisimov and B. Fitzharris, "Polar Regions (Arctic and Antarctic)," in contribution of Working Group II to the TAR of the IPCC, *Climate Change 2001: Impacts, Adaptation, and Vulnerability* (Cambridge: Cambridge University Press, 2001), 814, 831.

83   Susan Joy Hassol et al., *Impacts of a Warming Arctic: Arctic Climate Impact Assessment (ACIA)* (Cambridge: Cambridge University Press, 2004), 57.

84   Anisimov and Fitzharris, "Polar Regions," 831.

85   Raymond A. Rogers, "The Usury Debate, the Sustainability Debate, and the Call for a Moral Economy," *Ecological Economics* 35 (2000).

86   R. Rogers, *Solving History: The Challenge of Environmental Activism* (Montreal: Black Rose Books, 1998), 60.

87   Nelson, *Reaching for Heaven on Earth*, 305.

88   Livingston, *Arctic Oil*, 144.

89   George Wenzel, *Animal Rights, Human Rights: Ecology, Economy and Ideology in the Canadian Arctic* (London: Belhaven Press, 1991), 159.

90   Ibid., 8.

91   Michele Stephen, *A'aisa's Gifts: A Study of Magic and the Self* (Berkeley: University of California Press, 1995), 321.

92   Ibid.

93 Ibid., 326.

94 Anne Primavesi, *Gaia and Climate Change: A Theology of Gift Events* (London: Routledge/Taylor & Francis Group, 2009), 133.

95 Sarah McFarland Taylor, *Green Sisters: A Spiritual Ecology* (Cambridge: Harvard University Press, 2007).

96 Ibid., 113.

97 Primavesi, *Gaia and Climate Change*, 17.

98 Ibid., 34–35.

99 Cairney et al., "The Neurobehavioural Consequences of Petrol (Gasoline) Sniffing," 82.

100 Maggie Brady, *Heavy Metal: The Social Meaning of Petrol Sniffing in Australia* (Canberra: Aboriginal Studies Press, 1992), 22; Cairney et al., "The Neurobehavioural Consequences of Petrol (Gasoline) Sniffing"; Denov and Campbell, "Casualties of Aboriginal Displacement in Canada"; Bernard Segal, "The Inhalant Dilemma: A Theoretical Perspective," in F. Beauvais and J. E. Trimble, *Sociocultural Perspectives on Volatile Solvent Use* (New York: The Haworth Press, Inc., 1997).

101 John A. Livingston, *Rogue Primate: An Exploration of Human Domestication* (Toronto: Key Porter Books, 1994), 138.

102 Sierra Club of Canada, *No More Idling: California Standards Needed Now!* 19 October 2006, http://www.sierraclub.ca.

103 Peter Tertzakian, *A Thousand Barrels a Second: The Coming Oil Break Point and the Challenges Facing an Energy Dependent World* (New York: McGraw-Hill, 2006), 141.

104 Ibid.

105 M. Hessing, M. Howlett, and T. Summerville, *Canadian Natural Resource and Environmental Policy*, 2nd Edition (Vancouver: UBC Press, 2005).

AMERICAN FUEL FOR A
GLOBAL APOCALYPSE

At the 2005 United Nations Climate Change Conference in
Montreal, the Inuit Circumpolar Conference submitted a petition to the Inter-American Commission on Human Rights
that argued the United States is legally responsible for climate
changes that are destroying "the Arctic environment" and
"hunting-based economy of Inuit."[1] Though the United Nations
reported prior to the conference that Canada's emissions had
increased by 24 percent and the United States' by 13 percent between 1990 and 2003,[2] the Inuit Circumpolar Conference only
singled out the latter nation. The reason for this decision was
not only because the science indicates the United States is "the
largest emitter of greenhouse gases," but also because its Republican government had not joined "the international effort."
In contrast, Canada's Liberal government that preceded Prime
Minister Harper's 2006 Conservative election had ratified a
national commitment to the Kyoto Protocol and was, apparently, just performing badly. South of the border, a more overt
and dismissive denial of this international response and related
climate research was displayed from the moment George W.
Bush first publicly spoke about the issue as president in 2001:
"We will not do anything that harms our economy, because first
things first are the people who live in America."[3] It was because
of this unconscionable attitude and resulting policies that the
Inuit Circumpolar Conference claimed the United States bears
responsibility for the destructive impacts of *Sila*'s northern
warming.

The approach President George W. Bush took to climate
change had a familial history stretching back to when the Intergovernmental Panel on Climate Change began its interdisciplinary endeavour in the early 1990s. It was at this time that his

father, Republican President George H. W. Bush, inaugurated America's conservative approach to climate research by releasing a memorandum that proposed "the best way to publicly deal with concern about global warming would be to raise the many uncertainties."[4] Similar actions became prevalent in his son's government eight years later and then in Harper's Canadian Conservatives, even as the IPCC and *Arctic Climate Impact Assessment* documented a heightened scientific consensus.[5] In the first official post-election response to the question of the Kyoto Protocol and a policy for reducing fuel consumption, President George W. Bush's White House Press Secretary Ari Fleischer was emphatic:

> That's a big no. The President believes that it's an American way of life, and that it should be the goal of policymakers to protect the American way of life. The American way of life is a blessed one.[6]

Influencing this Republican conservation of the "American way" from what was portrayed as flawed climate policy and research was, as will be discussed, a unique blend of corporate fossil fuel and fundamental Christian interests.

While the energy sector has an obvious financial stake in delaying any reduction in fossil fuel consumption, Kevin Phillips explains some powerful Christian interests were also supporting the Republican government's delaying approach because climate change was seen as "irreconcilable with the Book of Genesis."[7] He clarifies that the economically conservative executives of energy and automotive industries did not have to believe in the end times of their Christian partners. Rather, there was simply an alignment between their concern with delaying climate and fuel policies and the politics of "the economically undemanding religious right."[8] Offering a related analysis that resonates with the wasteland economizing described in the previous chapter, Bill Moyers writes the energy sector's view of "the environment as ripe for the picking" and the fundamental Christians' regard of "the environment as fuel for the fire that is coming" coalesced in "President Bush's approach to the climate and environment."[9] For the Inuit represented by the Inuit Circumpolar Conference petition, this cultural defence was seen as having a global impact on the North that some climate research portrayed in terms as dark as the Christians. Even though the IPCC's most likely scenario has been the gradual continuation of warming over the next two centuries due to rising greenhouse gases, it has, with each successive report, recognized an increasing possibility of more abrupt projections that have an almost apocalyptic feel. This chapter not only examines the way in which the Republican alliance to conserve American culture has fuelled a dark sorcerous impact on the North, North America, and Gaia but also considers the potential adaptive value of apocalyptic beliefs for a global conscience.

CONSERVATIVE APOCALYPSE

While the road to the Inuit Circumpolar Conference's human rights petition may be directly linked to President Bush's 2001 rejection of the Kyoto Protocol, it can be indirectly related to the powerful historic hold fossil fuels have had on America's formal economy. In the opening decades of the twentieth century, the United States was the first nation to shift from coal to oil as a national strategy for energizing transportation. Analyzing the role automobiles have played in this cultural tradition, Morris Berman writes that by 1910 America was "the world's foremost automobile culture, with nearly a half million cars registered."[10] Within two decades, there were 5.3 million American vehicles and a 1933 "President's Research Committee on Social Trends reported the existence of an automobile psychology" that was making the population car-dependent.[11] Paired with these mental changes were environmental changes to the land as governments spent $245 billion on highways between

1947 and 1970. The more than half-a-billion cars at the beginning of the twenty-first century fuelled a jump in oil production from twenty to 3000 million metric tons since 1900.[12] Comparable to these American trends, the average twenty-first-century Canadian is projected to consume 575,000 litres of crude oil and drive 700,000 kilometres—the equivalent of travelling the planet's equator 17.5 times. [13] Since this American and, by association, Canadian cultural way has emitted "three times more CO_2 per person per year than Europeans and over a hundred times more than the citizens of the least developed countries,"[14] its continued denial of climate research and policy can be seen, as proposed by the Inuit Circumpolar Conference petition, as responsible for *Sila*'s northern warming and Gaia's climate changes.

The period between 1990 and 2002 not only saw oil and transportation industries contribute $415 million to America's Republican and Democrat politicians but also saw them lobby to ensure that tax breaks for oil continued and that the auto industry benefitted from weakened fuel-efficiency standards.[15] While these initiatives transcend party politics, the Republican administration of George W. Bush was so deeply enmeshed with the energy sector that some have argued it is difficult to delineate where one ended and the other began.[16] The resulting policies were reflected in a 2003 budget that reduced alternative energy research and gave "billions in subsidies for oil, gas, coal and nuclear energy," followed by a 2005 Energy Bill that sidelined climate change language and issues of energy efficiency in favour of increasing fossil fuel exploration and production.[17] This tendency began in 2001 when Vice President Dick Cheney's National Energy Policy, which was largely informed by energy executives, advocated more fossil fuel exploration, production, and consumption as the main solution to immanent energy shortages not seen "since the oil embargoes of the 1970s."[18] Central to the Republican administration's defence of the American way of life would be Canada's Alberta tar sands, what Vice President Cheney described as "a pillar of sustained North American energy and economic security."[19]

Considering Canada's frontier economy tradition and its expanding post-World War II economic ties with the United States, such a conservative approach had some well-positioned Canadian advocates. On July 14, 2006, Prime Minister Harper made it clear in a presentation to the Canadian and United Kingdom chambers of commerce that his government supported the Republican vision of the tar sands. He stated that with Canada's oil production forecasted to reach about four million barrels a day by 2015, of which two-thirds will come from the tar sands, and Canada being "a stable, reliable producer in a volatile, unpredictable world," it is no wonder that American policy-makers and investors "now talk about Canada and continental energy security in the same breath."[20] The prime minister would continue to globally promote an image of Canada as an emerging energy superpower. Meanwhile south of the border, the United States Congress declared in a 2006 report that "the proximity of this growing source of supply is a highly positive development for the U.S. and indeed the world."[21] Such a glowing assessment by American and Canadian political leaders was obviously based upon a highly sceptical view of the relation between fossil fuels and climate change, not to mention great faith in formal frontier economics.

The Republican energy policy drew its support from public opinion that was attached to an automotive tradition, uncertain about climate change, and generally thought a response should not entail extensive change—characteristics that could equally describe the Canadian public. In the early 1980s, public surveys on climate change found only a small fraction of the one-third who were aware of the subject recognized it "was mainly due to carbon dioxide from fossil fuels."[22] By the end of the century, surveys found half the public were aware of climate change, but their knowledge was generally inaccurate and unsure about the role of fossil fuels.[23] Many viewed climate change as a function of human greed and corruption, a moral decline that had become irreversible because of the need for so many people to change. When it came to action, respondents were largely apathetic and either thought an eventual technical fix would save the day or, in an almost Christian sense, that apocalyptic collapse would right the moral decline. Though the Christian tradition that influenced this moral interpretation of climate change began to find a political home in the Republican Party in the 1970s, it was not until George W. Bush's election that a Republican administration overtly aligned with the Christian fundamentalists by appointing them to government posts and international delegations.[24] The first term House Majority Leader was Tom DeLay, an influential fundamentalist who used his position to state: "Only Christianity offers a way to live in response to the realities that we find in this world."[25] Prior to President Bush's 2004 re-election, journalist Bill Moyers noted that Christian fundamentalists backed 231 legislators and gave 80 to 100 percent approval ratings to 45 senators and 186 members of Congress.[26] With this Christian support, Bush's Republican administration actively defended the cultural drive behind America's twentieth-century prosperity.

It is in this context of powerful political economic interests and popular support that a president from oil-rich Texas was elected who described in his autobiography, *A Charge to Keep*, an ascent to political power based upon a Christian "foundation that will not shift."[27] This statement is quite consistent with a Christian fundamentalist view that, as in Islamic and Judaic versions, tends to hold an uncompromising belief in the divine authority of its actions, works politically to bring about its vision of social change, and utilizes technology and social organizations consistent with propagating its mission.[28] Research on presidential speeches and policies led ethicist Peter Singer to write that beyond beliefs that interweave fundamentalist Christianity with close ties to the energy sector, the Bush administration appeared to "lack any clear and consistent philosophical underpinning."[29] Bush's speeches up until 2003 referred to good and evil 30 percent of the time, with Singer explaining that the use of "evil" was as "a *thing*, or a force, something that has a real existence" rather than "an adjective to describe what people do."[30] After documenting similar trends during the president's first term, Moyers wrote he had "no idea what President Bush thinks of the fundamentalists' fantastical theology," but did find their influence to be reflected in the images and metaphors he used.[31]

In facilitating an alliance with the fossil fuel sector, President Bush provided the fundamentalists with power to influence climate policy and expand their critique on what they saw as an immoral environmentalism that included climate research.[32] This fundamentalist view was epitomized in televangelist Pat Robertson's 1991 bestselling book, *New World Order*, in which environmentalists are described as evil priests of a pagan religion serving a godless liberal order.[33] Following his assertion that Christians need to actively prepare the world for God's revelation, fundamentalists have critiqued United Nations environmental events, like the 1997 Kyoto and 2005 Montreal meetings, as espousing anti-American and anti-Christian views that worship the planet based on "pseudoscience."[34] It is with such interpretations that Christians expanded their international activism against a godless environmentalism by utilizing both nongovernmental forums and the Republican government to manifest their "vision of social order"[35]—one that had some troubling dark dimensions.

In 2002 a poll found 59 percent of Americans "believe that the prophecies found in the Book of Revelation are going to come true."[36] Drawing from another poll that indicated 53 percent of the population believe Jesus' return and accompanying apocalypse are imminent, Singer writes, "we need to remember that tens of millions of Americans hold an apocalyptic view of the world."[37] In President Bush's America, there were varying fundamentalist Christian interpretations of this apocalyptic belief system. While dispensationalists literally interpret an apocalyptic chronology onto the immediate future that requires little action on their part, reconstructionists promote a more active missionary role in Christianizing America and the world as part of the divine plan. One view held in common by these passive and active wings is an interpretation of environmental crises as "portents of the Rapture, when born-again Christians, living and dead, will be taken up into heaven."[38] Considering the popularity of this belief, it is not surprising that many surveyed about climate change in 1999 tended "to move the topic from global warming itself to more familiar topics, such as moral deterioration."[39]

A recent American crisis with fossil fuel and climate dimensions displayed the interpretive power of this apocalyptic Christian belief. On August 29, 2005, Hurricane Katrina hit New Orleans and the state of Louisiana. As each day revealed unprecedented damage and social dislocation, many voices came forward offering interpretations of why it happened. Amidst scientific and policy arguments about inadequate levees, a city below sea level, and climate change projections of more powerful and frequent hurricane events, there arose Robertson's apocalyptic interpretation of this sign. While broadcasting *The 700 Club* on September 4, 2005, he connected Hurricane Katrina to an American moral failure that was nationally represented in the *Emmy Awards'* choice of New Orleans-born Ellen Degeneres, "an avowed lesbian," to be host.[40] In his words, "Is it any surprise that the Almighty chose to strike at Miss Degeneres' hometown?" After *Dateline Hollywood* misinterpreted his theological position, Robertson wrote a letter that clarified "Hurricane Katrina occurred because New Orleans is the epicenter of sinful jazz music" and that Ellen Degeneres merely epitomizes the city's liberal immorality.[41] Offering a similar view was Repent America's report that Hurricane Katrina had ended the city's sinful behaviour, with its director, Michael

Marcavage, asking that "this act of God cause us all to think about what we tolerate in our city limits, and bring us trembling before the throne of Almighty God."[42] From this perspective, a disruption like Hurricane Katrina is not due to inadequate levees, poor urban planning, greenhouse gas emissions, or the power of fossil fuels but rather reflects God's disapproval of a sinful city that liberally supports immoral beliefs and practices.

Though climate research cannot directly connect a single extreme event like Hurricane Katrina to climate change—but can connect it to a pattern that is consistent with projections—it was not because of such scientific uncertainty that these Christians rejected its interpretations. Rather, climate research based on a planet millions of years older than the six-thousand-year history represented in the Bible's Genesis story is the primary reason. Analyzing the historic evolution of apocalyptic beliefs, Arthur Mendel writes that the rejection of worldly "sinners as too corrupt for repair and their condemnation to total destruction" is this belief's first principle.[43] Because people are corrupted, arguments from the other side, such as liberals or climate researchers, are considered meaningless except as signs that the end is close. Clarifying this point, Stephen O'Leary explains the apocalyptic prophecy that many will reject salvation means the "argument becomes a mode of ritual enactment that retraces the pattern of the divine revelation."[44] Combined with this ritualization of apocalyptic belief is a second principle that absolutely believes in the immanent onset of apocalyptic justice and planetary transmutation.[45] Those who have been converted to this immovable foundation have access to privileged interpretations of events like Hurricane Katrina as "being part of a predetermined endtimes scenario orchestrated by a moral God."[46] A transcendent certainty overlays the wasteland's increasing uncertainty, thus freeing oneself from any personal, political economic, or cultural sense of responsibility for fuelling these catastrophes. Considering the evidence connecting climate change to fossil fuel consumption, one can argue that a similar transcendent belief also informs American car culture and Canadian frontier economics.

In response to the popular hold Christian fundamentalists like Robertson have had on America's religious imagination, environmental theorist Bill McKibben argues its theology is largely un-Christian.[47] His proposal is based upon a number of American surveys that reveal only 40 percent of the population know more than four of the Ten Commandments, about half can name one of the four Gospel authors, and 75 percent "believe the Bible teaches 'God helps those who help themselves.'"[48] The latter point suggests many believe the American idea of formal economic self-interest is Christian when it actually derives from Ben Franklin.[49] While McKibben argues the American public has replaced the Bible's "deep sharing and personal sacrifice" with a competing creed of economic self-interest[50]— as manifested in the Christian fundamentalist's Republican alliance with the fossil fuel sector—the previous chapter's elucidation of a wasteland theology suggests this American Christianity partakes in a tradition that has long influenced today's economizing mind. It seems that not only is a fundamentalist apocalyptic view on an extreme event like Hurricane Katrina based on the denial of climate research, Inuit knowledge, and international policy, but is also based on the marginalization of discordant Biblical messages that promote economic rather than religious ends. Such a limited knowledge base for those in positions of global power, like President Bush, could have some apocalyptic potential for peoples as far removed from the United States as the Inuit.

Beyond the fundamentalist authority of a Robertson or the political language of a President Bush, there is a deeper evolutionary reason that the belief in apocalypse has held so much power on the American and, more broadly, human mind. It is, after all, a belief that predates the current American incarnation in European Christianity and pre-Christian religious traditions from the Middle East and Egypt,[51] not to mention many other unique cultural guises the world over.[52] Drawing upon anthropological research, Daniel Wojcik proposes the popularity of these beliefs is based on the common experience of disruptions whose uncertainty lead people to explain them in the form of supernatural forces.[53] Of a similar view, O'Leary explains that apocalyptic beliefs are cultural strategies for understanding and responding to the uncertainties in people's lives.[54] What makes this belief seem so irrational to a more secular-minded liberal position is its continual reinterpretation of social and environmental disruptions as revelatory of a final and universal 'End' that never historically arrives.[55]

This historic mistake seems to be circumvented in the two apocalyptic views found in many indigenous North American cultures. Oral stories from prior to Christian contact either refer to a moral cosmos that is regionally—not globally or universally—responsive to human actions, or to earthly cycles of destruction and renewal that mirror "the cycles of the seasons with its endless circle of degeneration, death, and rebirth."[56] A short digression into the ancestral roots of an Inuit view on recurring apocalypses that have climate change dimensions will allow us to begin reinterpreting the above Christian view in relation to a contemporary global conscience.

NORTHERN APOCALYPSE

During Knud Rasmussen's 1929 Fifth Thule meetings, a man by the name of Tuglik told the ethnographer the following story of Inuit cosmological origins:

> There was once a world before this, and in it lived people who were not of our tribe. But the pillars of the earth collapsed, and all was destroyed. And the world was emptiness. Then two men grew from a hummock of earth. They were born and fully grown all at once. A magic song changed one

of them into a woman, and they had children. These were our earliest forefathers, and from them all the lands were peopled.[57]

It is a story that seems to tell of Inuit cultural emergence from the contact between the Thule, the latest pre-colonial culture to arrive in Alaska from Siberia at around 1000 C.E., and the Dorset Palaeo-Eskimo culture that by this time had already called the North, from Alaska to Greenland, home for some 3,500 years.[58] Interestingly, both Inuit stories and Western interdisciplinary research suggest *Sila*'s past climate changes played a significant role in apocalyptically transforming the North's pre-Inuit cultural pillars.

In Robert McGhee's research of Inuit origins, the Dorset's disappearance is reconstructed from archaeological evidence, oral stories, and conjecture based upon various historical forces that are commonly related to cultural change. The Dorset's way of living, which had evolved during a period of cooler temperatures than present, began dealing with a challenge starting around 700 C.E. as the medieval warming brought its changes.[59] This warming initiated a series of northern environmental and cultural disruptions that resulted in a mixture of dangers and opportunities for Dorset hunters, who, over a single lifetime, saw places that had rarely seen a caribou "become a summer home to tens of thousands."[60] Just as with today's northern melting, coastal areas that were frozen year-round suddenly became the summer home for migrating walrus, narwhal, and beluga. On land, the warmth resulted in the North American treeline shifting north some sixty miles. These changes would have brought unpredictability to a Dorset indigenous knowledge based on cooler tundra ecologies, probably bringing "starvation conditions among small communities, forcing members to move in order to find new sources of food or to seek help from neighbouring bands."[61] It also resulted in the Dorset coming into contact with more southerly-based cultures that were following the North's warming *Sila* and melting Northwest Passage from two different directions.

Long before Canada's colonial missions to the Inuit North, the Norse culture of northwestern Europe began colonizing the North Atlantic and initiated Western contact with these Inuit ancestors. In 982, Erik the Red found traces of inhabitation on Greenland that McGhee proposes must be the Dorset. Eleventh century Norse chronicles continue to document meetings along the large island's eastern coast, and archaeological evidence of smelted sheet copper at Dorset sites suggests an "intensive relationship between the two peoples."[62] Around 1012, a Norse chronicle refers to a meeting with Dorset in the Labrador, Canada, area that began as a trading relationship and soon deteriorated into a fight that saw a number of Dorset killed.[63] These relations with the West seem to have continued, as Venetian trader Nicolo Zeno reported spending part of 1395 at a monastery in Greenland that was using the inherited Inuit kayak and "round domed stone houses" that mimicked the igloo.[64] Such intercultural adaptations signify the dawn of a Western approach to the North that long before today's interest in *Inuit Qaujimatuqangit*'s (IQ's) traditional ecological knowledge continued in the fifteenth century with Breton, Norman, and Basque whalers adapting the Inuit harpoons to their "distinctive whaling technology,"[65] and later in the nineteenth century with British explorers documenting the value of Inuit culture in relation to northern clothing, tools, survival techniques, and, most importantly, guides who were "perceptive readers of snow and ice conditions, weather patterns, and the behaviour of animals."[66] With the cooling temperatures of the fifteenth and sixteenth centuries, Western knowledge of the earlier northern relations gradually receded as many Nordic Christians abandoned their Greenland settlements. For those who stayed, their communication with Europe eventually went silent, with historians speculating the settlements were destroyed by a combination of conflict with Inuit ancestors, disease, and drought.

Unlike the Norse, whose northwestern movements toward Greenland were impeded by their commitment to European agricultural adaptations in a time of cooling, the Thule culture coming eastward toward the Dorset from the Asian continent had a different adaptive approach to the North. As the earlier medieval warming melted the Northwest Passage, the bowhead whale population shifted northward above the North American continent, with the Thule following. Employing "efficient whaling techniques," iron technologies based on East Asian trade networks, and boats that allowed them to travel "several hundred kilometers" in a summer season, the Thule created semi-permanent coastal settlements anywhere a whale kill occurred between Alaska and Greenland.[67] During the ice-locked northern winters, the Thule used their whaling technology with dog-drawn sleds to secure inland resources. With these powerful weapons, McGhee explains this Asian culture was successful in most encounters with the Dorset when violence was required to secure sustenance. The Thule were also successful because they unintentionally brought from Asia infectious diseases like influenza that had the same result for the Dorset as the virgin soil epidemics that impacted the Inuit during colonial contact with the West.[68] Within a few centuries, the Dorset were reduced to small pockets. By the second wave of European explorers that began arriving in the sixteenth century looking for a Northwest Passage, the distinct three-thousand-year-old Dorset culture seemed to have apocalyptically disappeared from the North.[69] The "pillars of the earth" had collapsed for the Dorset, but, according to Tuglik's story, their culture would inform the genesis of the Inuit.

While the Norse receded southward with the cooling Little Ice Age that began in the fourteenth and intensified in the sixteenth century, the dwindling numbers of Dorset became important partners for a Thule culture dealing with a freezing Northwest Passage and a southward shifting whale population. Based on Inuit oral stories that speak of "stolen wives and adopted orphans," McGhee writes that "minor elements of Dorset language, culture and view of the world may have been incorporated" by the more dominant Thule.[70] Both the igloo and the harpoon technology of the Inuit are suspected to have derived from this Dorset influence.[71] As the open water and whaling opportunities receded, the Thule became more regionally adapted to a "wider range of habitats" by following the Dorset approach of being more dependent on smaller animals like seals, caribou, and fish.[72] With regard to spiritual understandings, it is thought that the Dorset came to inform the shamanistic religion of personal spirit helpers, while the Thule offered many of the common practices and cosmological beings that transcend Inuit cultural differences across the North. Overall, the once economically richer and uniform Thule culture was gradually transformed into a diversity of Inuit groups like the Polar Eskimo of Greenland, Central Inuit of Baffin Island, and Caribou Inuit of western Hudson Bay.[73] It seems the cooling brought an apocalyptic collapse to the Thule's pillars as well, resulting in a myriad of evolving Dorset-Thule or Inuit cultures with unique regionalized substantive ways, IQs, and *Silatuniq* for northern living.

Supporting Jaypeetee Arnakak's dynamic definition of a more-than-traditional *Inuit Qaujimaningit* in the Introduction, these pre-industrial changes in the North led Wenzel to conclude the position of environmentalists like Livingston during the seal hunt debates have been based upon a very limited historic conception of what constitutes Inuit culture. The blending of "Thule technology and maritime specialization with Dorset environmental knowledge" reveals that for at least one thousand years Inuit culture and IQ has adjusted to climate and resource changes by utilizing new technologies and social adaptations.[74] This adaptive view was further highlighted by Arnakak when we corresponded about the Thule-Dorset influence in relation to the conflict between Western research and indigenous knowledge on the Asian cultural origins of indigenous people. Drawing upon his present experience as a starting point, he stated that for a future anthropologist it would appear "Inuit culture died out in the 1960s (and perhaps it did), but I still consider myself an Inuk."[75] After clarifying that the term "Inuk" would not have been used by the Dorset or Thule, Arnakak added that despite his present self-identification as Inuit, "my ancestors say we've been here since the beginning."[76]

The cooling temperatures and shifting ecologies may have facilitated a more meaningful, perhaps even complementary, relation between the Dorset and Thule in the creation of the Inuit. Recognizing this possibility does not necessitate denying power dynamics related to population size, political economics, and technological adaptations, but rather simply suggests a deeper syncretism of the Dorset and Thule in making the Inuit, as Tuglik said, "fully grown all at once." As such, this story not only contrasts Western research's continuing marginalization of IQ, President Bush's fundamentalist conservation of American culture, and Canada's continuation of a frontier economy in the tar sands, but also hints at the kind of dynamic intercultural dialogue that is required if we are to conscionably respond to today's sorcerous climate changes.

To further appreciate the IQ sense of ancestry, Arnakak explains it is necessary to recognize that in Inuit creation stories time is not linear but is rather "more like a person slowly achieving cognition and memory in a way that things fall into place."[77] Long before the remembrance of Dorset-Thule influence on the Inuit, there comes a more distant and obscure memory of cosmological origins that was told to Rasmussen by Naalungiaq:

> In the very first times there was no light on earth. Everything was in darkness, the lands could not be seen, the animals could not be seen. And still, both people and animals lived on the earth, but there was no difference between them ... A person could become an animal, and an animal could be-

come a human being ... That is the way they lived on earth in the very earliest times, times that no one can understand now.[78]

After this darkness come the stories of collapsing pillars, Inuit origin during the Little Ice Age, and colonial contact at the dawn of today's industrial warming. Prior to Canada's early-twentieth-century colonial missions are storied memories of a nineteenth-century return to the Thule whaling way with Scottish and American whalers in search of whale oil.[79] It was only as America and Canada shifted in the early twentieth century to an oil-based energy strategy that Inuit memories of these whaling relations gave way to the Canadian missions discussed in the previous chapter. This IQ-means of remembering offers a way of interconnecting the time when "everything was in darkness" to those more recent struggles of the Inuit with both environmental researchers at the Canadian Polar Bear Roundtable and a Republican government aimed at conserving American culture.

While the conversion of Inuit to Christianity may have been hastened by twentieth-century Canadian missions, like that of the Grey Sisters in Chesterfield Inlet, the first sign of the approaching Christian message was heard by many Inuit in masses on American whaler boats.[80] As in the deep ancestral past, the Inuit found something of value in this religion that many connected with their substantive ways of living, IQ, and *Silatuniq*. Exemplifying this approach, Bernie Putulik stated during the workshop that "I believe in the Bible and in my culture." This is not to say all Inuit take such an approach. As Arnakak related to me, the Christian conversions were so successful that some Inuit "would rather see Inuktitut and IQ die, and do actively pursue that end."[81] Others from Chesterfield Inlet supported Putulik's more flexible approach in their description of a contemporary Inuit apocalyptic vision of today's warming. Many participants drew Christian inspiration from the stories of Noah and the prophetic Revelations, with Elizabeth Tautu highlighting Chapter 24 of the Gospel of Matthew as a means "to understand what is happening with climate change." In this passage, Jesus' disciples ask him to explain what he meant when he said all the stones of the temple would be overturned. After describing wars, famines, earthquakes, and an evil that makes people's love "grow cold," Jesus explains that "the end will come" only when the word of God is "preached through all the world for a witness to all mankind."[82] As with America's fundamentalist Christian interpretation of events like Hurricane Katrina, there was for many in Chesterfield Inlet a resonance between this story and the North's contemporary climate changes. Such an apocalyptic view of the present situation is not limited to the religiously minded, for many Western researchers of a more liberal rationality are painting a similarly dark possibility for the future of the Canadian North, American south, and Gaia in general.

CLIMATIC EPIC

In 2003 a controversial United States Pentagon briefing on the potential impacts of climate change received global media attention. Rather than projecting the more common assumption of a gradual warming, the report by Schwartz and Randall used paleoclimate research to model an alternative future with an apocalyptic feel.[83] Up until 2010, this model depicts warming as an "economic nuisance" due to storms, droughts, and heat waves, but then things change abruptly as the North Atlantic Ocean's thermohaline circulation begins to reverse because of the influx of freshwater from melting northern and southern glaciers.[84] This ocean circulation currently moderates Western European temperatures by acting like a conveyor belt that sinks the Atlantic's highly dense cold sea water as it follows North America's coast toward the equator before cycling the warm surface water, air, and energy in a northeasterly direction. As with paleoclimate research that indicates past glacial changes to this pumping process resulted in an unexpected northern cooling during a naturally occurring global warming cycle, their model projects a sudden cooling of up to 14°C, drops in rainfall, droughts, and agricultural scarcities for "Northern Europe and eastern North America."[85] During the first five years of these Atlantic changes, "the effects are far more pronounced in Northern Europe," but after 2015 the chill becomes harsher for southern Europe and much of North America. Adding concern are the projected global impacts of increasing social instability as regional and national conflicts are intensified due to energy, food, and water scarcities. Their security picture for America is one of a nation turning "inward, committing its resources to feeding its own population, shoring-up its borders, and managing the increasing global tension."[86] It is a dark vision that resonates with the Thule-Dorset challenge, as well as the apocalyptic beliefs of some American and Inuit Christians.

One mitigating factor in Schwartz and Randall's model of climate-induced social conflict is a demand for oil that they project will increase by 66 percent over the next three decades and thus outstrip supply—even with the intensification of tar sands production.[87] The research of ecologist Jared Diamond in his book, *Collapse*, similarly argues that while oil and gas will be accessible for a few more decades, these reserves will increasingly "be deeper underground, dirtier, increasingly expensive to extract or process."[88] While more pessimistic proponents of such "peak oil" projections suggest the remaining oil is at about a trillion barrels and thus place the peak at about 2010,[89] sceptics of this position push the peak back. But as Paul Roberts argues, the problem here is that the peak is a plateau followed by a 2016 cliff. Even the conservative energy policy that came out of Republican Vice President Dick Cheney's National Energy Policy group was ultimately built on the fear of "facing catastrophic energy shortages."[90] At either the extreme of liberal research or conservative denial, Roberts states the end is still relatively "imminent, given the size and value of the oil-based infrastructure—the tankers, the pipelines, the refineries, 747s, Greyhound buses, and, above all, cars—that would need to be upgraded or replaced."[91] Considering these infrastructural and political economic challenges, he concludes that "the real question is not whether oil is going to run out (it will) but whether we have the capacity, the political will, to *see* the outcome soon enough to prepare ourselves for it."[92] This is a relevant question not only for Americans but for Canadians who have seen their Conservative government deny the need for a climate response while promoting the unsustainable extraction of the tar sands at an ever quickening pace.

The extensive investment required to convert the tar sands into crude oil has been described by Prime Minister Harper as "an enterprise of epic proportions, akin to the building of the pyramids or China's Great Wall. Only bigger."[93] Contrasting such an optimistic, perhaps even delusional, vision is Andrew Nikiforuk's description of the tar sands as not only "the world's dirtiest hydrocarbon" but also "what a desperate civilization mines after [it has] depleted its cheap oil."[94] The United States Department of Energy can be seen to support such a critique with figures that estimate on average one barrel of oil can "pump out anywhere between twenty and sixty barrels of cheap oil," while the same amount only yields between four and five barrels in the tar sands.[95] Beyond being reflective of peak oil realities, the tar sands' open-pit mining operations also evoke for Nikiforuk hellish images that bespeak of apocalypse. As he writes, "companies must mow down hundreds of trees, roll up acres of soil, drain wetlands, dig up four tons of earth to secure two tons of bituminous sand, and then give those two tons a hot wash."[96] While politicians like Harper, Cheney, and Bush describe these developments in positively "epic" terms, Nikiforuk echoes the analysis of Roberts by concluding the tar sands "distracts North Americans from two stark realities: we are running out of cheap oil, and seventeen million North Americans run their cars on [it]."[97] Considering the interrelation between peak oil and climate change, I propose there is a very different epic of apocalyptic proportions facing Americans, Canadians, Westerners, and the world. Returning once more to the words of Nikiforuk, the destructiveness of the tar sands "should be a bold invitation for us to live within our means, exercise prudence, and abandon the oil-fuelled mythology of consumption without limits."[98]

Offering a historic perspective on this apocalyptic challenge to American and, by association, Canadian culture, Kevin Phillips draws lessons from the declines of the British Empire at the turn of the twentieth century and the Dutch in the eighteenth century.[99] His research suggests these successive national downturns in political economic fortunes were related to an inability to adopt more productive energy options that became available as wind gave way to coal in the nineteenth century and coal to oil in the United States' twentieth century. The early twenty-first century's unique reality is not only that there is no energy source more productive than fossil fuels, but also that the challenge is coming in the form of both climate change and peak oil projections. Despite these historic differences, the five symptoms common to past national declines are, in Phillips's view, consistent with the direction in which President Bush led America: "widespread public concern over cultural and economic decay," growing interest in moralistic interpretations, rising commitment to faith paired with a marginalization of science, a popular embrace of apocalyptic interpretations of present realities, and "hubris-driven national strategic and military overreach."[100] In 2005 he saw the darkening signs of the future in the dire forecasts of the speculative housing credit bubble, ever-increasing foreign debt, oil scarcities, and increasing climate disruptions. Looking over the coinciding 2010 to 2020 timeframes of these issues, Phillips writes that he "can't remember anything like this multiplicity of reasonably serious calculations and warnings," adding if only "one or two of the four are correct, major troubles lie ahead."[101]

Such extreme visions of North American cultural collapse have resonances with the research of Jared Diamond and Joseph Tainter on the role of climate change in past civilizational collapses in places like Easter Island, Mexico, and elsewhere.[102] The recurrence of social collapses are for these thinkers great historic dilemmas because they represent the cultural difficulty of adapting to the "rapid loss (within two or three generations) of an established level of social, political, and economic complexity."[103] Some common factors that interconnect past collapses include population growth and the adoption of regionally damaging environmental practices that, in Diamond's words, result in "food shortages, starvation, wars among too many people fighting for too few resources and overthrows of governing elites by disillusioned masses."[104] These past changes were often not gradual, but rather the decline occurred rapidly after reaching a cultural peak in power that probably made the quick change a shocking surprise to most.

As with the climatically collapsing pillars of the Thule-Dorset experience, a common factor highlighted by Tainter and Diamond in past collapses was regional warming or cooling trends that challenged societal stability. These changes did not bring societal collapse on their own, but, as Tainter explains, were most destructive when the "perturbation came after a period of declining marginal returns to investment in complexity."[105] In other words, societal collapses related to climate changes are more likely when internal inflexibility and external conflicts due to environmental scarcities—like peak oil—are already affecting cultural adaptability. This is also what the abrupt climate model of Schwartz and Randall projects as America and the world potentially revert to a pre-Industrial "norm of constant battles for diminishing resources."[106] These are epic visions, though clearly quite different from Harper's view on the tar sands.

While Diamond agrees with much of Tainter's research, he also criticizes him for reasoning today's "complex societies are not likely to allow themselves to collapse through failure to manage" the environment.[107] In contrast, Diamond proposes the central point of past collapses is that complex societies can fail because of inadequate ecological management, despite the appearance of great social power. Based on this view, Diamond identifies four types of cultural failures: 1) the inability "to anticipate a problem"; 2) the problem arrives and there is a failure "to perceive it"; 3) once it is perceived "they may fail even to try to solve it"; and, 4) the solution "may not succeed."[108] Analyzing the Bush administration's environmental record, Diamond concludes that under his leadership, the United States was enacting the third failure of not acting despite perception. But the Republican alliance of fossil fuel interests with Christian fundamentalists suggests Diamond may not have properly assessed this failed response. President Bush's selective fundamental reading of the Bible and rejection of contradictory climate research appears exemplary of Diamond's second—not third—type of failure, such that the interconnected symptoms of climate change and peak oil are not being perceived because of the powerful cultural influence of fossil fuels on North American ways of living. A similar critique can be levelled at a Canadian wasteland approach to the tar sands.

Denying climate research and policy was President Bush's approach up until July 2005 when, at the G8 Summit in Scotland, he at least agreed human activities are linked to climate change. Later that month he signed an Asian-Pacific climate agreement with Australia, China, India, South Korea, and Japan that was a more palatable alternative to the Kyoto Protocol because, as with Prime Minister Harper's Clean Air Act, the focus is on developing clean energy technology and omitting greenhouse gas reduction targets.[109] The ability to maintain this conservative position in the face of extensive contradictory evidence was related to the Bush administration's powerful oil-Christian alliance and the willing compliance of a population dependent upon this energy source. As Roberts argues, questions concerning the role of energy companies in America's energy and security policy remain muted—even while the nation was mired in the Iraq war—because such soul-searching might force individuals to see themselves "as extensions of an out-of-control energy system that begins at home, in our own cars and houses."[110] The same can be said of Canadian ambivalence about the relation between tar sands production and the nation's political denial of climate change research and policy.

Looking at today's challenging energy situation, environmental historian Alfred Crosby concludes that we are either "standing on the peak of our energy achievements poised for the next quantum leap," or are teetering toward Gaia's "standard operational procedure of pairing a population explosion with a population crash."[111] Adding a religious dimension to Crosby's assessment, Glenn Scherer muses that "with the speed of climate change now seen as moving much faster, global warming could very well be a major factor" in fulfilling Christianity's apocalyptic prophecies.[112] The rational projection of such sorcerous possibilities suggests President Bush's and Prime Minister Harper's conservative delays have made the eventual transformation of America and Canada more costly and difficult, while also leaving increasingly less space in this collapsing bottleneck for those who are more liberally minded to enact a global conscience.

CONCLUSION: APOCALYPTIC *SILATUNIQ*

Though one alternative to Diamond's assessment is that America's apocalypse is due to the maladaptive perception of a fossil fuel lifestyle, there is a more irrational and destructive possibility. If Christian fundamentalists actually informed President Bush's vision of America and were not simply a political ploy for votes, then his inadequate climate policy could be seen as supporting the arguments of John Livingston, Lynn White, Jr., and Primavesi on the pivotal role of Christianity in today's crisis. This is not only because this tradition has politically promoted a fundamental rejection of climate research and policy, but also because of its historic religious influence on the West's continuing formal economization of a Gaian wasteland. In an interesting turn of events that supports the Inuit Circumpolar Conference human rights petition, the attempt to conserve North American culture is fuelling climatic signs like Hurricane Katrina, retreating glaciers, shifting ocean circulation, warming *Sila*, and odd polar bear behaviour—all of which herald the fundamentalist's Second Coming of Jesus. Based on a religious study of the current social and environmental situation, both O'Leary and Keller propose that "Western civilization has been acting out a self-fulfilling prophecy."[113] Only Diamond's first category of failing to anticipate the problem characterizes such a maladaptive religious influence on today's sorcerous climate changes, but to make such an assessment requires a broader interdisciplinary approach than that which is offered by him or much of the climate research. His analysis in fact highlights the marginalization of religion in liberalized research and political economics that is as problematic as the selective Biblical reading and scientific denial reflected in President Bush's conservative Christian views of Gaia's climate changes and *Sila*'s northern warming.

While fundamentalist Christians like Pat Robertson supported President Bush's denial, there was also arising in the United States an alternative Christian view on the moral responsibility required to meet this spiritual challenge that could broaden and strengthen the liberal political approach. In 2002 the Evangelical Environment Network's What Would Jesus Drive campaign promoted transportation choices as "moral

choices for three basic reasons: impacts of transportation pollution on human health, particularly that of children; the problem of global warming and its impacts on the poor; and the
consequences of our oil dependence on national and economic
security."[114] This was followed in 2006 by eighty-six evangelical
leaders signing onto the Statement of the Evangelical Climate
Initiative and its assertion that we "now believe that the evidence demands action."[115] Contrasting Robertson's Genesis-
based denial of climate research, the statement combines climate research and Genesis passages to propose that humanity
is failing to act as good stewards and this situation "constitutes a
critical opportunity for us to do better."[116] It consequently supports various moral actions like praying to God for awareness,
studying "the Bible in light of the impacts of global warming,"
reducing energy use, purchasing more fuel-efficient vehicles,
becoming more reliant on renewable energy, and writing government officials and the media. Such action is deemed important not only because "any damage that we do to God's world is
an offense against God," but also because marginal populations
will be hit hardest and the Bible tells us "to protect and care for
the least of these as though each was Jesus Christ."[117] It is an
approach that highlights White's point about the importance of
religion being due not only to its maladaptive potential, but also
that "religion would need to be part of the solution."[118]

This religious call "to do better" is also very much in line with
the spirit of the Inuit Circumpolar Conference's human rights
petition, and further highlights the need for climate research
and politics to broaden its interdisciplinary and intercultural
considerations during this time of potentially apocalyptic cultural change. The religious wisdom of shifting our cultural approach to climate change was further intimated in an August
4, 2006 broadcast of *The 700 Club* that saw Robertson tell his
viewers that the record-breaking heat wave then blanketing
the United States was converting him to the view, "If we are
contributing to the destruction of the planet we need to man-

age it."[119] His subsequent embrace of the Evangelical Climate Initiative reveals the potential of localized changes to call forth a view on apocalypse as a practice rather than as a worldly end. Though Robertson's warming experience may have revealed to him some of climate research's value, his universalizing fundamentalism probably still limits the extent to which he could engage discordant interdisciplinary and intercultural research.

Providing a deeper sense of what apocalypse can offer conservative and liberal climate views on adaptation, O'Leary explains the historic prevalence of apocalyptic signs and cultural interpretations is an important pattern that reveals "the End has been misunderstood."[120] Apocalypse can be a global cultural end that occurs at some specific moment in time, but on a more practical everyday level it can inform an adaptive ethical practice for facing a future that is continually "breaking in and exploding every complacency." In other words, historically recurring climate changes are a "normative standard against which our actions may be measured."[121] Without such an apocalyptic sensibility, it is likely impossible for America, Canada, and, more broadly, the West to fulfill Tainter's call to "become the first society in history to recognize the processes by which problem-solving abilities decline and to devise corrective actions."[122] The Inuit manifestation out of the apocalyptic Thule-Dorset experience suggests this new challenge does in fact have cultural precursors, and in the next chapter we will further consider this possibility by returning to the *Silatuniq* of an Inuit view on a northern apocalypse of Gaian scale.

ENDNOTES

1   Inuit Circumpolar Conference, "Inuit Petition Inter-American Commission on Human Rights to Oppose Climate Change Caused by the United States of America, 7 December 2005, http://www.inuitcircumpolar.com.
2   United Nations Framework Convention on Climate Change, *Key GHG Data*, 2005, http://unfccc.int.2860.php.
3   Quoted in Peter Singer, *The President of Good & Evil: The Ethics of George W. Bush* (New York: Dutton 2004), 135.
4   Spencer Weart, *The Discovery of Global Warming* (Cambridge: Harvard University Press, 2003), 168.
5   For example, see Robert F. Kennedy, *Crimes Against Nature: How George W. Bush and His Corporate Pals Are Plundering the Country and High-Jacking Our Democracy* (New York: HarperCollins, 2004); Weart, *The Discovery of Global Warming*.
6   Singer, *The President of Good & Evil*, 135.
7   Kevin P. Phillips, *American Theocracy: The Peril and Politics of Radical Religion, Oil, and Borrowed Money in the 21st Century* (New York: Viking, 2006), 366.
8   Ibid., 67.
9   Bill Moyers, "Welcome to Doomsday," *The New York Review of Books*, 24 March 2005, 52, 5.
10   Morris Berman, *Dark Ages America: The Final Phase of Empire* (New York: W. W. Norton and Company, 2006), 54.
11   Ibid., 54.
12   John Robert McNeill, *Something New under the Sun: An Environmental History of the Twentieth-Century World* (New York: W. W. Norton and Company, 2000).
13   David R. Boyd and David Suzuki Foundation, *Sustainability within a Generation: A New Vision for Canada* (Vancouver: David Suzuki Foundation, 2004), 5.
14   Tim F. Flannery, *The Weather Makers: How We Are*

Changing the Climate and What It Means for Life on Earth
(Toronto: HarperCollins Canada, 2006), 299.

15   Phillips, *American Theocracy*, 44.

16   Andrew Austin and Laurel Phoenix, "The Neoconservative Assault on the Earth: The Environmental Imperialism of the Bush Administration," *Capitalism Nature Socialism* 16, no. 2 (2005).

17   Linda McQuaig, *It's the Crude, Dude: War, Big Oil and the Fight for the Planet* (Toronto: Doubleday Canada, 2004), 303.

18   Cited in Andrew Nikiforuk, *Tar Sands: Dirty Oil and the Future of a Continent* (Vancouver: GreyStone Books, 2008), 30.

19   Cited in ibid., 31.

20   Prime Minister Stephen Harper, "Harper's Index: Stephen Harper Introduces the Tar Sands Issue," 14 July 2006, http://www.dominionpaper.ca/articles/1491.

21   Cited in Nikiforuk, *Tar Sands*, 32.

22   Weart, *The Discovery of Global Warming*, 117.

23   John Immerwahr, *Waiting for a Signal: Public Attitudes Toward Global Warming, the Environment, and Geophysical Research* (New York: Public Agenda, 1999), 2, http://Earth.agu.org/sci_soc.html.

24   See Kennedy, *Crimes Against Nature*; Joel A. Carpenter, *Revive Us Again: The Reawakening of American Fundamentalism* (New York: Oxford University Press, 1997); Sara Diamond, *Spiritual Warfare: The Politics of the Christian Right* (Montreal: Black Rose Books, 1990); Doris Buss and Didi Herman, *Globalizing Family Values: The Christian Right in International Politics* (Minneapolis: University of Minnesota Press, 2003).

25   Quoted in Singer, *The President of Good & Evil*, 110.

26   Moyers, "Welcome to Doomsday."

27   Quoted in Singer, *The President of Good & Evil*, 97.

28   Richard T. Antoun, *Understanding Fundamentalism: Christian, Islamic, and Jewish Movements* (Walnut Creek, CA: AltaMira Press, 2001).

29   Quoted in Singer, *The President of Good & Evil*, 89.

30   Ibid., 2.

31   Moyers, "Welcome to Doomsday."

32   Buss and Herman, *Globalizing Family Values*, 20.

33   Pat Robertson, *The New World Order* (Dallas: Word Pub., 1991), 227; also see Kennedy, *Crimes Against Nature*.

34   Buss and Herman, *Globalizing Family Values*, 22.

35   Ibid., 136.

36   Cited in Glenn Scherer, "The Godly Must be Crazy: Christian-Right Views Are Swaying Politicians and Threatening the Environment," *Grist Magazine: Environmental News and Commentary*, 27 October 2004, http://www.grist.org/news/maindish/2004/10/27/scherer-christian.

37   Singer, *The President of Good & Evil*, 208.

38   Scherer, "The Godly Must be Crazy."

39   Immerwahr, *Waiting for a Signal*, 13.

40   Quoted in *Dateline Hollywood*, "Robertson Blames Hurricane on Choice of Ellen Degeneres to Host Emmys," 5 September 2005, http://datelinehollywood.com/archives/2005/09/05/robertson-blames-hurricane-on-choice-of-ellen-degeneres-to-host-emmys/.

41   Pat Robertson, "Letter: Pat Robertson Corrects Dateline Hollywood Article," 18 September 2005, http://datelinehollywood.com/archives/2005/09/18/pat-robertson-corrects-dateline-hollywood-article/.

42   Repent America, "Hurricane Katrina Destroys New Orleans Days before 'Southern Decadence,'" 31 August 2005, http://www.repentamerica.com/pr_hurricanekatrina.html.

43   Arthur P. Mendel, *Vision and Violence* (Ann Arbor: University of Michigan Press, 1999), 42.

44   Stephen D. O'Leary, *Arguing the Apocalypse: A Theory of Millennial Rhetoric* (New York: Oxford University Press, 1994), 205.

45   Mendel, *Vision and Violence*, 42.

46   Daniel Wójcik, *The End of the World As We Know It: Faith,*

Fatalism, and Apocalypse in America (New York: New York University Press, 1997); Catherine Keller, *Apocalypse Now and Then: A Feminist Guide to the End of the World* (Boston: Beacon Press, 1996), 55.

47 Bill McKibben, "The Christian Paradox: How a Faithful Nation Gets Jesus Wrong," *Harper's Magazine*, August 2005, 31–37.

48 Ibid., 31.

49 Ibid.

50 Ibid., 33.

51 For example, see Norman Cohn, *Cosmos, Chaos and the World to Come: The Ancient Roots of Apocalyptic Faith* (New Haven: Yale University Press, 1993).

52 For example, see Mircea Eliade, *The Myth of the Eternal Return: Or, Cosmos and History* (Princeton, NJ: Princeton University Press, 1971).

53 Wójcik, *The End of the World As We Know It*, 54.

54 O'Leary, *Arguing the Apocalypse*, 26.

55 Carpenter, *Revive Us Again*; Wójcik, *The End of the World As We Know It*; Keller, *Apocalypse Now and Then*; O'Leary, *Arguing the Apocalypse*.

56 Clara Sue Kidwell, Homer Noley, and George E. Tinker, *A Native American Theology* (Maryknoll, NY: Orbis Books, 2001), 153–154.

57 Quoted in John Bennett and Susan Diana Mary Rowley, *Uqalurait: An Oral History of Nunavut* (Montreal: McGill-Queen's University Press, 2004), 161.

58 Richard G. Scott et al., "Physical Anthropology of the Arctic," in M. Nuttall and T. V. Callaghan, *The Arctic: Environment, People, Policy* (Amsterdam: Harwood Academic Publishers, 2000).

59 Robert McGhee, *Ancient People of the Arctic* (Vancouver: UBC Press, 1996).

60 Ibid., 197.

61 Ibid.

62   Ibid., 194.

63   E. W. Nuffield, *The Discovery of Canada* (Vancouver: Haro Books, 1996).

64   Quoted in Nuffield, *The Discovery of Canada*, 39–40.

65   J. R. Miller, *Skyscrapers Hide the Heavens: A History of Indian-White Relations in Canada* (Toronto: University of Toronto Press, 2000), 285.

66   Ibid.

67   McGhee, *Ancient People of the Arctic*, 184.

68   Ibid., 223.

69   Ibid.

70   McGhee, *Ancient People of the Arctic*, 233; also see Miller, *Skyscrapers Hide the Heavens*, 284.

71   McGhee, *Ancient People of the Arctic*, 233.

72   Ibid., 231.

73   Miller, *Skyscrapers Hide the Heavens*; McGhee, *Ancient People of the Arctic*; Barry Lopez, *Arctic Dreams: Imagination and Desire in a Northern Landscape* (Toronto: Bantam Books, 1989).

74   George Wenzel, *Animal Rights, Human Rights: Ecology, Economy and Ideology in the Canadian Arctic* (London: Belhaven Press, 1991), 27.

75   E-mail correspondence, 6 March 2006.

76   Ibid.

77   E-mail correspondence, 6 March 2006.

78   Quoted in Bennett and Rowley, *Uqalurait*, 161.

79   Miller, *Skyscrapers Hide the Heavens*.

80   *Amarok's Song: Journey to Nunavut*, directed by Martin Kreelak and Ole Gjerstad, Words and Pictures Video, in association with the National Filmboard of Canada and the Inuit Broadcasting Corporation, 1998.

81   E-mail correspondence, 30 March 2005.

82   *Good News Bible: The Bible in Today's English Version* (Toronto: Canadian Bible Society, 1976), Mt. 24:5–14.

83   Peter Schwartz and Doug Randall, *An Abrupt Climate*

Change Scenario and Its Implications for United States National Security (Washington, DC: Pentagon, 2003).

84 Ibid., 8.

85 Ibid., 9.

86 Ibid., 13.

87 Ibid.

88 Jared M. Diamond, *Collapse: How Societies Choose to Fail or Succeed* (New York: Viking, 2005), 490.

89 Paul Roberts, *The End of Oil: On the Edge of a Perilous New World* (Boston: Houghton, 2004), 52.

90 Ibid., 52.

91 Ibid., 101–102.

92 Ibid., 65.

93 Prime Minister Stephen Harper, "Harper's Index: Stephen Harper Introduces the Tar Sands Issue," 14 July 2006, http://www.dominionpaper.ca/articles/1491.

94 Nikiforuk, *Tar Sands*, 16.

95 Ibid., 15.

96 Ibid., 13.

97 Ibid., 16.

98 Ibid., 179.

99 Phillips, *American Theocracy*, 11.

100 Ibid., 220.

101 Ibid., 95.

102 Joseph A. Tainter, "Global Change, History, and Sustainability," in R. J. McIntosh, J. A. Tainter, and S. K. McIntosh, eds., *The Way the Wind Blows: Climate, History, and Human Action* (New York: Columbia University Press, 2000); Joseph A. Tainter, *The Collapse of Complex Societies* (Cambridge: Cambridge University Press, 1988); Diamond, *Collapse*.

103 Tainter, "Global Change, History, and Sustainability," 332.

104 Diamond, *Collapse*, 6.

105 Tainter, "Global Change, History, and Sustainability," 350.

106 Schwartz and Randall, *An Abrupt Climate Change Scenario*, 16.

107 Diamond, *Collapse*, 420.

108 Ibid., 421.

109 Jane Perlez, "U.S. to Join China and India in Climate Pact," *New York Times*, 27 July 2005, http://www.nytimes.com.

110 Roberts, *The End of Oil*, 304.

111 Alfred W. Crosby, *Children of the Sun: A History of Humanity's Unappeasable Appetite for Energy* (New York: W. W. Norton, 2006), 164.

112 Scherer, "The Godly Must be Crazy."

113 O'Leary, *Arguing the Apocalypse*, 220; Keller, *Apocalypse Now and Then*, 12.

114 National Association of Evangelicals, *For the Health of the Nation: An Evangelical Call to Civic Responsibility*, http://www.nae.net (accessed 16 July 2006); also see Statement of the Evangelical Climate Initiative, *Climate Change: An Evangelical Call to Action*, 2006, http://www.christiansandclimate.org.

115 Statement of the Evangelical Climate Initiative, *Climate Change*.

116 Statement of the Evangelical Climate Initiative, *Climate Change*, refers to Gen. 1:26–28.

117 Statement of the Evangelical Climate Initiative, *Climate Change*, refers to Mt. 22:34–40; Mt. 7:12; Mt. 25:31–46.

118 Gary Gardner, "Engaging Religion in the Quest for a Sustainable World," in L. Starke, ed., *State of the World 2003* (New York: W. W. Norton and Company, 2003), 161.

119 Bruce Wilson, "Pat Robertson's Sweaty Global Warming Epiphany Challenges American Environmental Movement," 5 August 2006, http://www.talk2action.org; World Net Daily, "A Green Gospel: Pat Robertson Converts—to 'Global Warming,'" 3 August 2006, http://www.WorldNetDaily.com.

120 O'Leary, *Arguing the Apocalypse*, 219.

121 Ibid., 219, 223.

122 Tainter, "Global Change, History, and Sustainability," 349.

MAKING CARBON
CONFESSIONS TO *SEDNA*

After coming in as the runner-up for the 2007 Nobel Peace Prize to the Intergovernmental Panel on Climate Change (IPCC) and former Vice President Al Gore, the once Inuit Circumpolar Council Chair Sheila Watt-Cloutier stated "the issue has won and, in fact, our own planet Earth was a winner in all this."[1] The Nobel committee explained their choice of the IPCC and Gore as being due to their tireless work "to disseminate greater knowledge about man-made climate change."[2] While the IPCC was documenting climate research, Gore had, in the last years of the George W. Bush presidency, offered a liberal reading of an unfolding climate apocalypse in the popular documentary, *An Inconvenient Truth*. With the changes occurring much quicker than had been projected in the 1980s and 90s, he argues that America's, and, more broadly, humanity's fossil fuel dependence will need to undergo a swift and radical change. By making homes more energy-efficient, increasing transportation efficiency, employing more renewables, and promoting carbon dioxide sequestration technology, Gore proposes that America could quickly reduce its emissions to below 1970 levels without significantly affecting affluence.[3]

Though Gore's proposals could bring significant change, the lack of analysis on how culture and religion are part of the problem or solution reflects a liberal resistance to recognizing the limits of rationality. As I argued in the previous chapter's critique of Diamond and climate research in general, this inability to appreciate the potential adaptive role of culture and religion in assessing Gaia's indeterminate dimensions will continue to narrow our interdisciplinary and intercultural research. What we need is something more in line with the spirit of Gore's Nobel Prize acceptance: "We have to quickly find a way to change the world's consciousness about what exactly we're facing, and why we have to work to solve it."[4] To conceive how we are to enact such an apocalyptic change in consciousness, this final chapter returns once more to an Inuit view on relating with Gaia's animated northern ecology.

When the elder Naalungiaq told Knud Rasmussen of the darkness before the Dorset-Thule world, he also talked of a being known as *Sedna* that he said is "the most feared of all spirits, the most powerful, and the one who more than any other controls the destinies of men."[5] Stories of *Sedna* describe a watery being that can change the *Sila*, increase winds, bring forth a blizzard, and, most importantly, "make the animals disappear so that people go hungry" when offended by improper human actions.[6] Around today's Chesterfield Inlet, Her potentiality seems to be evoked in the declining arctic char population, northward migrating birds, and mammals from the southern treeline, and the recession of an ice ecology that has been home for polar bears and seal species. Though the climate research that informed Gore's analysis may not be inclined to a *Sedna* or Gaia view, it should by now not be surprising that the West's scientific knowledge of animal changes is consistent with *Inuit Qaujimatuqangit*'s (IQ's) ecological knowledge. The *Arctic Climate Impact Assessment* states northern wildlife is being impacted by habitat loss, increased risk to migrating diseases, rising pollution levels, difficult competitions with what were once more southerly species, and expanding human industrial developments.[7] Putting this northern research in a global context is another study of 1,700 species that found animals have, on average, shifted their range pole-ward by six kilometres every ten years and, more broadly, spring is moving forward 2.3 days per decade.[8] As the IPCC suggests, climate-forced migrations will further threaten Gaia's already receding biodiversity.[9] Inuit are expressing concern about the *irksina* changes to polar bear behaviour and the animals they hunt. Such regional changes would in pre-colonial times call forth stories of *Sedna* and the kind of consciousness needed for a response—stories that are still of importance to many Inuit despite the conversionary impacts of colonialism.

While *Sila*'s northern warming and *Sedna*'s shifting animals are calling forth many Inuit responses, the global scale of today's sorcerous changes highlight the point that Canadians and Westerners need to consider their own conscionable approach to this northern situation. Inspired by Gore's call for a new consciousness, in June 2008 Liberal opposition leader Stéphane Dion outlined his Canadian response called the "Green Shift Plan," which would allow the public "to do what is right—not what is easy—for our environment and our future."[10] His alternative to Prime Minister Harper's Conservatives was to shift Canada's tax structure, such that taxes would be cut "on those things we all want more of such as income, investment and innovation," while taxes would increase on "what we all want less of: pollution, greenhouse gas emissions and waste."[11] In a press statement following the release of Dion's plan, Harper used a double entendre that characterized this "bad policy" as a "new tax grab" that, if elected, would truly deserve its "green shift" name.[12] When the prime minister called the election a few months later, Canadians faced a distinct choice between the Conservative's fossil fuel-based denial and this renewed Liberal vision of a global conscience. In this chapter, we will consider Canada's electoral response to these political options in light of a traditional Inuit approach to *Sedna*'s potentially apocalyptic northern animal changes.

MOTHER OF THE ANIMALS

The origin story of *Sedna* begins with a young woman named Nuliajuk who angers her father by refusing to marry, despite being of an appropriate age to have children.[13] One evening a handsome man enters Nuliajuk's room and, after turning into a dog, has sex with her. Pregnant from the affair, Nuliajuk gives birth to half-human/half-dog children who are taken with her to live on an island. The dog-man husband regularly returns to her father to get food for the family, but the father gets tired of this arrangement and so plans an accident that drowns the dog-man. When the father subsequently brings food to the island, he is torn to pieces by the dog children, as requested by Nuliajuk. With nobody to bring them food and starvation looming, she desperately sends her children away in a watercraft. The now isolated Nuliajuk is tricked once more into a relationship with a handsome man who promises to take her from the island. This man offers "her a life of comfort," but when leaving "for his land across the sea" she realized her "betrothed was in fact a sea-bird."[14] With Nuliajuk crying on the seabird's kayak, her dead father miraculously returns to life and hears her anguish.[15] In the rescue attempt, he takes Nuliajuk onto his kayak and tries to escape from the seabird. Angered by this rejection, the bird dives at the kayak until the *Sila* becomes stormy and the sea turbulent. Terrified, the father throws Nuliajuk overboard to calm the seabird and elements, but she hangs onto the edge of his kayak. As she clings to its side, her father cuts "off her fingers at the first knuckle."[16] These "pieces fell into the sea and became ringed seals," but she managed to continue hanging on to the boat.[17] He then "cut all her fingers at the second joint, and those pieces swam away as bearded seals," while "other pieces of her fingers became walruses and whales."[18] With each finger joint becoming a different northern animal, she descended to the bottom of the sea and became the climatically influential *Sedna*, both "creation and creator."[19]

Though known across the North's various regions by names like Nuliajuk, Her most common name, *Sedna*, means "the one down" in the deep waters, as explained to me by Jaypeetee Arnakak. He added that in IQ it is considered improper etiquette to refer to an ancestor or elder with their proper name, and as such using the descriptive name *Sedna* reflects an attempt to

not offend.[20] Such an etiquette is particularly important because She is not simply an ancestor, but what Inuit describe as
an indweller that "lives at the bottom of the sea and controls
all the animals."[21] While Arnakak wrote that "an indweller is a
personification of a spiritual concept,"[22] Daniel Merkur characterizes indwellers as physically immanent numina that are
"more closely analogous to Western concepts of the forces of
gravity and magnetism than to concepts of angels, demons, and
God."[23] Offering another view, Mircea Eliade refers to *Sedna*
as the "source and matrix of All life," the one whose goodwill
Inuit are dependent upon.[24] Supporting these ethnographic
views, Arnakak writes of Her as the "Source of all creation and
destruction."[25]

In trying to help me transcend Western assumptions that
continued to limit my appreciation of *Sedna*, Arnakak asked
me to "view ecology, environment and wildlife as sentient beings that are deserving of your respect." This animism that
informs "IQ and other hunter-gatherer spiritualities" was
described by him as a deep level of thought that "points to a
real identification with and relatedness to reality and Nature's
wonderful cycles."[26] The difficulty that Westerners, including
myself, have historically had in appreciating this worldview is
a central concern of Nurit Bird-David's ethnographic research
with the Nayaka of South India.[27] Responding to the theories of Enlightenment thinkers like E. B. Tylor,[28] she explains
Western thought assumed the primacy of its own belief in an
economically self-interested individual that consequently led it
to view indigenous people as making the mistake of attribut-
ing a similar animistic self onto natural others. The result was
decades of research dedicated to defining the historic origin of
this "mistaken strategic guess."[29] For someone like Tylor, even
Christianity was conceived as developing out of this mistake,
and it was not until the Enlightenment that Westerners evolved
a science for rectifying this original error. The resulting view

that animism and science are "fundamentally antithetical" became one primary Enlightenment argument against the validity of indigenous knowledge like IQ.[30]

In contrast to the traditional Western view of animism as a mistaken personification, Bird-David's experiences with the Nayaka led her to conclude that people "do not first personify other entities and then socialize with them but personify them as, when, and because [they] socialize with them."[31] While Western objective research comes to knowledge through separating the self from some reduced focus of study, she found the Nayaka's "relational epistemology" was concerned with "developing the skills of being in-the-world with other things, making one's awareness of one's environment and one's self finer, broader, deeper, richer."[32] Such an approach seems remarkably similar to Arnakak's assertion that IQ's animism recognizes "Nature has the last word, and humanity's obligation is to reflect that *Silatuniq*."[33]

As with the Inuit view on an indwelling *Sedna*, the animistic cosmos of the Nayaka is not limited to humans, animals, and plants but also engages cosmological beings that are part of the local ecology known as *devaru*.[34] *Devaru* cannot be understood using the term "spirits" or "supernatural beings;" rather, Bird-David suggests thinking about them as "superpersons" who mediate relations between the Nayaka and the other beings of their ecology.[35] A hill *devaru*, for example, "objectifies Nayaka relationships with the hill; it makes known the relationship between Nayaka and that hill."[36] Such animated hunter-gatherer perspectives also have an interesting resonance with the eco-theology of Anne Primavesi, for she writes that being of Irish Catholic descent means her regional sense of Gaia is personified in the Gaelic landscape mother-goddess "known as the *Cailleach Bhéara* or Supernatural Female Elder."[37] Just as the Inuit connect *Sedna* with northern waters and the Nayaka associate *devaru* with socially important hills and other areas, Primavesi's *Cailleach Bhéara* is "attached to natural features of the physical landscape" that has a wildly fertile and untameable power.[38] This personified landscape figure is associated with the "flocks and herds of beast" throughout Ireland, and is locally "accessible to human consciousness and human communication" in certain animals and sacred places like mountains.[39]

Returning to Bird-David, she explains it is the process of maintaining relations with the beings of an ecological place that allows them to be seen not only as people, but also as "relatives" who share their very essence with the human community.[40] Contrasting a Canadian, American, and Western focus on maximizing economic self-interest in a scarce wasteland, the cultural traditions of Nayaka, Celts, and Inuit assume a social abundance in Gaia's diverse regions that requires of humans a particular etiquette.[41] It is in this sense that the ancestral *Sedna* can be conceived as the indwelling animator of the Hudson Bay's cold waters, coastal lands, and animals—an Inuit relative of regional proportions and capacities who speaks through predictable patterns and unexpected events like today's climate changes, warming *Sila*, and shifting polar bears.

The Inuit climate research of Shari Fox hints at this animistic sensibility in a discussion she had with the elder Aqqiaruq on *Sila*'s northern warming.[42] Unclear about the term *uggianaqtuq* that was used to refer to these changes, Fox asked Aqqiaruq to explain it. He stated:

> I am very close with my sister. Say I wasn't feeling myself one day and I went to go visit her. As soon as I walk in the room, or say something, she would know right away that something is wrong. She would ask me, "is there something wrong with you?" She would say I was *uggianaqtuq*. I was not myself, acting unexpectedly or in an unfamiliar way.[43]

Offering a similar assessment of the climate around Chesterfield Inlet, Simionie Sammurtok explained the *Sila* has become seasonally and daily unpredictable, stating "at the start of the day it is really calm and about an hour later it is different. It is off and on, off and on." Drawing upon IQ's historic depth by comparing stories of *Sila* from his grandfather's day in the late 1880s to today, Louie Autut added people want to use IQ but "everything is changing because of the climate change." As we saw in the Introduction and in the first chapter, these changes are challenging the capacity of IQ to adapt to the future. Making the *uggianaqtuq* of a changing *Sila* even more difficult today is its relation to the unpredictable changes of *Sedna's* animated Hudson Bay waters and related animal movements.

By far the most talked about animal concern in Chesterfield Inlet was the observed changes in *Sedna's* arctic char population. Over the course of the workshop, many spoke about the char as an important source of income and diet that is being impacted by warming waters. As Elizabeth Tautu explained:

> Some fish are dying out from going upriver, the river is too low, the lakes are low, some ponds are empty, there is no water because of the climate change. The *Sila* seems to get warmer every year. There are lots of rivers around this area, and lots of fish try to go upriver and they are dying out because of the low water.

Concurring, Louie Autut stated the water has become so low "you can just take them out of the river without a spear or fishing line," and Andre Tautu added, "we can cross rivers with runners that in past seasons we would not even be able to cross with hip-waders." Many expressed concern that the change in spawning time from mid-September to late-August, coupled with lowered water, was impacting their harvests as increasing numbers of fish die or are bruised while going up river. Such IQ observations have been supported by the *Arctic Climate Impact Assessment's* projection that warming waters will create less "optimal thermal conditions" for char by shrinking potential habitat, providing conditions for invasive species from warmer waters, as well as increasing their respiration and thus uptake of pollutants.[44]

The char are not the only northern animal expressing an *uggianaqtuq* that is animating *Sedna*'s waters with various uncertainties. As with the changed seasonal timing of the char runs, Andre Tautu talked of the delay in water fowl egg-laying, stating "there was no time for them to lay the eggs because of the changing *Sila*." The IPCC projects that marine and tundra birds will be impacted by the effects of warming on their migration patterns.[45] In contrast to past climate changes when species could adapt by migrating to appropriate habitats, today the situation is more challenging for long-range migrations because of the need to bypass "freeways, agricultural zones, industrial parks, military bases, and cities."[46] Birds that migrate great distances and require various seasonal habitats will be especially vulnerable to warming trends.[47] Migratory issues are not secluded to the sky, for caribou will have land difficulties related to permafrost melting and other terrestrial changes. Meanwhile, in the water, the ability of Inuit to hunt seals will be challenged by a reduction of ice that will make ice-dependent ringed, ribbon, and bearded seals more vulnerable, while harbour and grey seals from more temperate waters will extend their ecological niches.[48]

It is with these seal changes that concerns about the polar bear also migrate into *Sedna*'s story, for as we have seen, their hunting success depends "upon good spring ice conditions."[49] When this is not the case, then their hunger may very well result in increasing *irksina* experiences for Inuit in a context of general animal *uggianaqtuq*. Just as seals and polar bears traverse between the sea and the land, Rasmussen was informed that *Sedna* "is everywhere, not only out in the sea, but in the interior too, where she may suddenly emerge out of the ground or up out of the lakes."[50] As the elder Naalungiaq clarified, She is "mistress of everything else alive, the land beasts too, that mankind had to hunt."[51]

In Chesterfield Inlet, the polar bear was the most talked about land animal because the warming temperatures had increased their frequency around the community and hunting camps. These threats that were talked about in the Introduction were dramatized by Andre Tautu when he described his experience of leaning against a camp door to keep out a hungry bear. The changes to sea ice that are impacting ice-dependent seals may be forcing polar bears to, in the words of the *Arctic Climate Impact Assessment*, "adapt to a land-based summer lifestyle," although it projects competition with brown and grizzly bears, coupled with increasing human interactions, will further threaten the polar bears' future.[52] Supporting some of this research, Elizabeth Tautu stated, "we see the brown bear, it is coming," while Andre Tautu said he was out hunting a polar bear and instead "caught a grizzly." This was the first time the participants had seen a grizzly bear near Chesterfield Inlet. While the grizzly bears and grey seals come north with a climatically supportive ecology, the polar bears and ice-dependent seals will struggle to shift their ice-based adaptations to other ways of living in *Sedna*'s quickly changing ecology. These multiple variables, in combination with the different positions of Inuit, climate researchers, and Canadian politicians, are what can be seen as underlying the conflicting views represented at the Polar Bear Roundtable.

What Primavesi's research adds to this regional view on *Sedna*'s shifting animals is a sense of Gaia's ability to globally respond to human behaviour. Climate change is part of Gaia's animated response, but the planet's offence is further reflected in the massive global extinction of animals in which *Sedna*'s changes partake.[53] Since 1600 the world has seen "484 animals and 654 plants" go extinct, and much of that damage has occurred during the twentieth century when about 1 percent of the planet's birds and mammals disappeared. According to the IPCC, "25% of the world's mammals and 12% of birds are at a significant risk of global extinction,"[54] with projections indicating between a 30 and 50 percent disappearance of terrestrial species "in the next century or two."[55] Supporting Inuit observations, the IPCC states "there is high confidence that rapid climate change, in conjunction with other pressures, probably will cause many species that currently are classified as critically endangered to become extinct and several of those that are labelled endangered or vulnerable to become much rarer."[56] If this happens, "it will be the sixth greatest extinction ever in earth's history," and it will be a human-induced event that is "far faster than any previous one."[57] While the cause of previous mass extinctions is still inconclusive, environmental historian J. R. McNeill argues the current "cause is obvious: a rogue mammal's economic activity."[58] The sorcerous externalization of greenhouse gas emissions are evoking Gaia, *Sedna*, and other animated regions to respond with animal disturbances. During times such as these, Inuit hunters in the past knew it was vital to consciously attend the signs of what is happening and ask questions concerning why.

A CONSCIONABLE ETIQUETTE

The Chesterfield Inlet workshop was held in October, which Arnakak explained to me is the time of the year most often associated with *Sedna*:

> October is a time of scarcity, a time of transition from water to ice. It is a time of hunger and *Sedna* is connected because shamans would be earning their keep this time of year when the sea becomes too choppy to hunt sea mammals safely ... *Sedna* is a powerful psychological concept that maintains social order in the bare face of adversity, and is therefore imbued with dark and wet elements.[59]

He added there is no "formal connection between *Sedna* and October," but rather Her power is appreciated at this time of year because the *qavaq*, or safe ice to hunt sea mammals, is not present. Such a safe hunting platform is largely receding today as *Sedna* challenges Inuit subsistence. While the IPCC estimates between 33 and 57 percent of Inuit subsistence is based upon local harvesting, the Chesterfield Inlet participants told me that the high cost of store food means approximately 75 percent of their diet is based on country food. As Louis Autut clarified, "We get more country food because we use our money to buy gas, oil and parts for our snowmobile, and this is what we spend our money on." While a successful hunt traditionally reflected *Sedna*'s benevolence, David Pelly warns She could be "vindictive when traditional customs were neglected."[60] Her anger at offensive behaviour would be signified in storms that "prevent the hunters from going to sea" and in the absence of animals.[61] Such animal changes as those occurring today would raise concern because they may signify the disruption of an ancestral relation that is central to Inuit sustenance. The hunter's relation with *Sedna*'s animals can help us clarify not only the impact of Canada's frontier economy, America's car culture, and, more inclusively, the West's formal economy, but also some northern insights on what an alternative global conscience may look like.

Stories of *Sedna* are part of an Inuit hunting cosmology that, according to Arnakak, highlight the importance of IQ's hunting etiquette.[62] Reflecting on the source of this etiquette, the elder Ivaluardjuk told Rasmussen the "greatest peril of life lies in the fact that human food consists entirely of souls."[63] Because souls "do not perish with the body," they must "be propitiated lest they should revenge themselves on us for taking their bodies."[64] These words resonate with the polar bear hunt described by George Wenzel in the Introduction, for the slow approach of his guide was similarly based in the intent of being respectful to the prey's desire. As Wenzel writes, an animal will only participate in the food process when the hunter approaches "the animal with an attitude of respect" and an intent to share that which the animal generously provides.[65] Similarly, Pelly states the "Inuit hunter is not extracting from the environment but creating a bond between his people and their environment," with the bond being based on *Sedna*'s animals giving their being to humans.[66] Other hunter-gatherer researchers, like Bird-David and Marshall Sahlins, have likewise found that close daily contact with a surrounding ecology leads to sharing relationships that have economic and spiritual dimensions.[67] With the human and ecological communities seen as sharing partners, such hunting cosmologies contrast the West's wasteland economics by implying "that as human agents appropriate their shares they secure further sharing."[68] Rather than Gaia and *Sedna* being primarily scarce, they are seen in this animated view as an abundant sharing cosmos in which humans need to participate.

Clarifying the "substantive" economic dimensions of this Inuit spirituality, Wenzel documents *ningiqtuq* as an Inuktitut reference to a web of social mechanisms for ensuring food and other resources are distributed from the individual to extended families and the broader community.[69] As with Arnakak's contextualizing definition of IQ, *ningiqtuq* primarily describes a social rather than economic process, such that the greatest amount of sharing occurs within the immediate social context of the extended family before radiating out to less immediate relations through activities like communal meals. The hunter, Zachariasie Aqiaruq, details such a sharing protocol following a walrus hunt:

> The ones that made the kill got the fore-flipper section and the ones that did not make the kill received the chest section. Those who came in afterwards would get the hind flipper section. The hunters that made the kill would get the areas that had more meat in them as their share of the catch.[70]

This sharing protocol reproduces the "giving cosmos," thus reminding humans they "are part of a transcendent universe in which everything emanates from the same spiritual source."[71] In contrast, improper hunting protocols—bragging about a kill, talking while eating or not sharing the catch—could, as the elder Ahlooloo reminisced, display a "lack of respect for the animal spirit world" and cause hunger.[72] If respect for the animals is not forthcoming, then *Sedna* could make the animals recede from the Inuit and thus bring great danger like that which is being experienced today with *Sila's* northern warming.

Prior to researching the colonial *ilira* and polar bear *irksina* of the Inuit, Hugh Brody spent time with the Beaver Indians of northwest Canada and relates a hunting story that can help us contemplate the relation of IQ's sharing etiquette to *Sedna* and, more broadly, the change of consciousness introduced by Gore. While out with a hunting party, Brody became bewildered by the constantly shifting consensus concerning where game might be found. He learned there were subtle variables that were not hard and firm, but rather required hunters to "accept the interconnection of all possible factors, and avoid the mistake of seeking rationally to focus on any one consideration."[73] There seemed to be no differentiation between theory and practice, as every decision remained alterable in light of changing circumstances.[74] Whereas plans and management strategies constitute "a decision about the right procedure or action," Brody found that during a hunt "there is no space left for a 'plan,' only for a bundle of open-ended and nonrational possibilities."[75] To choose the most fortuitous possibility, he came to realize that Inuit and other indigenous people often utilized dreams as an intuitive skill for condensing "data that is both voluminous and elusive," thus allowing hunters to attend more information than rationally possible.[76] As he further explains, on the hunt there are often conflicting signs, such that "wind changes urge a delay," while shifting animal movements require more speed.[77] Weighing difficult factors impresses upon one the indeterminate nature of knowledge, thus highlighting the need for dreams and other imaginative practices in directing our conscious decisions.

It needs to be clarified that the imaginative practice of

dreaming or entering *Sila*'s "great loneliness," as discussed in the first chapter, is not done in lieu of knowledge but rather is an additional intuitive practice for, in Brody's words, "paying the closest and deepest possible attention to the world."[78] In my discussions with Arnakak, he explained Inuit have various techniques to facilitate such intuitive sensibilities for a successful hunt:

> My father and grandfather always worked and puttered around with quiet but incessant humming to achieve a sustained low-level trance-state. The solutions, pictures, patterns, relations, come of their own accord to the receptive mind. Inuit women in the old days had this technique whereby solutions to difficult technical problems in sewing and design were constructed through their dreams.[79]

He also related that while the idea of meditating in a room is intriguing to him, the practice did not have the same effect as the "trance-like state" attained when "fishing through the ice,"[80] combining "hunting, rhythmic sound, and symbolic talismans" or listening to *Sedna* stories.[81] These words intertwine with Brody's to offer a view on the Hudson Bay waters, lands, and animals as *Sedna*'s changing thoughts, and the Inuit hunter as one who requires a certain cultural etiquette and method for consciously communing with this regional manifestation of Gaia.

While a deep awareness and proper etiquette can allow the hunter to successfully track an animal, it is with the following act of killing and eating souls that *Sedna*'s world of greater danger can appear. As Mark Nuttall states, Inuit hunters recognize they have a responsibility to ensure "animals are killed properly and their meat, bones and hide utilized in ways that will not offend the animal's guardian spirit."[82] To facilitate this offering of respect, IQ uses elaborate rituals performed prior to the hunt and after the animal's death.[83] For example, the elder Analok explained that before cutting up the seal, "you would get some meltwater from your mouth and pour it into the seal's snout," for it is believed "that a seal refreshed in this manner would be more likely to return again in the form of another seal for another drink."[84] A review of the world's diverse hunting rituals led anthropologist Roy Rappaport to describe "death and killing" as powerful "signs" that are ritually used to produce "trance-like states" capable of bringing to awareness many co-existent meanings.[85] The Inuit ritual of giving a dead seal a drink of water is clearly in line with this practice, as is Igjugarjuk's description of how a hunter tries to not anger *Sedna* while killing Her animals:

> Nothing is lost; the blood and entrails must be covered up after a caribou has been killed. So we see that life is endless. Only we do not know in what form we shall reappear after death.[86]

For Rappaport, such hunting rituals point to the limitations of our rational depictions, and thus "constitute attempts to push past representation in all its forms to naked, immediate existence."[87] As with hunting dreams and shamanic initiations, these hunting rituals are meant to foster a consciousness that respects the limited human capacity to symbolically conceive *Sedna*'s and Gaia's indeterminate changes.

Informing this respectful approach are what Arnakak defines as "an elaborate set of taboo systems we call *tirigusuusiit*."[88] He explained to me that these taboos are "ritualized conditions and obligations that kept in balance the social, ecological and cosmological order." Representing the cosmological dimensions of this aim for balance, Naalungiaq informed Rasmussen that most taboos are directed at *Sedna* because of the power She holds over humanity. In a review of hunting culture taboos, Rappaport similarly concludes they are rules about appropriate and inappropriate actions constituted in a kind of cosmological hierarchy. As he explains:

> At the apogee of this hierarchy stand a limited number of postulates concerning spirits ... Cosmological structure is elaborated in a second class, a class of axioms by or through which the spirits postulated are associated with elements and relations of the material and social world ... These relations are given greater concreteness and further specificity in yet a third level of understandings constituted of rules and taboos concerning action appropriate or inappropriate in terms of the understandings of the cosmological structure which inform them.[89]

This dissection of cosmology and taboo resonates with Inuit views of *Sedna*, for She is at the apogee of these stories that relate Her indwelling ecological changes and rules to human behaviour. As such, taboos reflect a cultural etiquette that recognizes the cosmological limits of human knowledge and actions, and the need for intuitive practices, rituals, and *tirigusuusiit* that can facilitate awareness of the ever-changing surround. During those unfortunate times when a hunter or community had breeched this etiquette, Inuit would then traditionally engage the specialized *Silatuniq* of a shamanic guide who could communicate with *Sedna*. It is with this part of the story that we can begin to more fully connect this animated northern etiquette to a contemporary global conscience.

When animals have been disrespected by unconscious hunters who have broken *tirigusuusiit*, the hair of *Sedna* is then often described as becoming full of "foggy revenge-seeking ghosts."[90] As Pelly relates:

> It was said that she was most happy when her hair was tidy, but she had no fingers with which to comb it. So when her

hair became figuratively tangled with the misdemeanours of Inuit, and her anger led to a decline in the number of her animals released to the hunters, the shaman had to pay a visit to her dwelling on the sea bottom. He had to comb her hair and thereby pacify the Mother of the Sea so that she would release the seals.[91]

A "retired shaman" in John Houston's documentary, *Nuliajuk*, similarly explained that when *Sedna* "inflicts hardships on us, when She withholds Her seals, Her thick hair falls down, holding back the sea beasts."[92] It was during such times of "desperate hunger, terrible storms, a world out of balance" that, he added, "only a shaman could set matters right."[93] Offering a similar view, Arnakak explains that *Sedna* is ultimately an "Inuit Shamanist God" that is meditatively engaged to gain insight on a situation. As we have seen in relation to shamanic and hunter rituals, he states IQ believes in the capacity of the "imagination to solve or transcend almost any problem."[94] In such a view, *Sedna*'s animal disappearances traditionally call the community to engage rituals that can determine any potential human sources of Her regional anger. It is a ritual worth reconsidering today as the offence of *Sedna* and Gaia manifests in shifting polar bears, seals, char, and many other northern and global animal changes.

At one such community séance, the shaman is described as beginning the ritual by breathing deeply and in silence for a time as his helping spirits were summoned.[95] Once this happened, the shaman began to murmur: "The way is made ready for me; the way opens before me!" In response, the audience answered in chorus: "Let it be so." It was then that the earth opened and the shaman struggled "for a long time with unknown forces." After bypassing these forces, he cried to the gathering that the way to *Sedna* was open. The group exclaimed: "Let the way be open before him; let there be way for him." At this point, the shaman began to cry "Halala-he-he-he, Halala-he-he-he" in an ever-receding voice that signified he had set off.

On the sea bottom far from the gathering, the shaman confronted obstacles to *Sedna* that represented Her anger at human indiscretions. The first was "three great stones in constant motion barring his road" that, after they were passed, led along a path to Her house on a hill. Then *Sedna*'s anger manifested a great wall in front of the house that the shaman needed "to knock down with his shoulder." With this obstacle passed, Merkur describes the meeting:

> At last the shaman finds the Sea Mother, seated with her back to her lamp and to the animals that are gathered around it. These are signs of her anger. As well, her hair is filthy with the misdeeds of Inuit and hangs loose and dishevelled over one side of her face. The shaman takes her by the shoulder and turns her face toward the lamp. He strokes, smoothes, and cleans her hair, as she, lacking fingers, cannot do. In this manner, he attempts to persuade her to release the animals.[96]

It is then that the shaman told *Sedna* there are no more seals, to which She answered in a spirit language: "The secret miscarriages of the women and breaches of taboo in eating boiled meat bar the way for the animals."[97] After summoning all his power to appease Her anger, the shaman returned to the séance and, following a long silence, demanded all to "confess their miscarriage of their breeches of taboos and repent." This communal confession to *Sedna* aimed to restore balance—a conscionable act that I argue desperately needs to be revisioned today in light of Gaia's smoggy greenhouse gas hair.

At the time of colonial contact, Inuit confessions for breeched *tirigusuusiit* were often related to women's menstruation, food preparation, and the already discussed hunting rituals. While many contemporary Inuit may not desire a return to these specific *tirigusuusiit*, Arnakak indicates that perhaps new ones are needed to deal with contemporary realities. Their definition would, for him, be based upon an intercultural sharing that we have seen is central to the Inuit Circumpolar Council's international advocacy and, more generally, to IQ:

> With our great knowledge, scientific insights and technology we should be able to come up with similar *tirigusuusiit*. The development of these living rules should be the measure of our collective wisdom. We can't do this through force, legislation and policy, it has to start by beginning to treat each other like thinking reasonable beings capable of taking on necessary and fair social and ecological obligations and responsibilities.[98]

After talking about the cultural impacts of colonialism in relation to *Sedna*'s animals, the Inuit elder Mariano Aupilarjuk similarly requested that the "Inuit system of laws and principles" be studied and understood "because in the future the land will need protection, the sky and sea will require care, and we would all then accept the responsibility."[99] These calls for reconceiving *tirigusuusiit* are consistent with Rappaport's elucidation of a fourth level in the cosmological hierarchy discussed above. Below taboos is a level that imports "material and social conditions immediately prevailing in the everyday world" into the hunting culture's cosmology, ritual, and everyday etiquette.[100] It is through this connective tissue that IQ's *tirigusuusiit* and ritual responses to *Sedna* can be continually re-interpreted in light of changing ecological realities—not unlike the Thule-Dorset marriage during their time of climatic apocalypse.

A contemporary reinterpretation of *tirigusuusiit* is exactly what Arnakak seems to be doing when he describes the empathy, humility, awareness, and sharing of a merciful hunt as the "matured traits of a socialization process that is diametrically opposed to the individualistic formal economy."[101] Rather than a wasteland scarcity, Arnakak describes IQ as a knowledge that "works with nature to produce something spiritual." Offering a similar critical perspective, Brody contrasts an indigenous hunting etiquette that maintains awareness of the "constant movements of nature, spirits, and human moods" with the planning, control, and management common to Western environmental research and policy.[102] The implication is that the West's universalizing rationality, as practiced by nations like Canada, has historically breeched *tirigusuusiit*, and continues to do so in its current response to Gaia's climate changes and *Sedna*'s animal disappearances. This is a different way of thinking about Canadian, American, and Western political economic dynamics, which began to impact Inuit with the colonial onset of frontier missions and are today manifested in *Sila*'s warming. Here, the conscionable act of confessing is redefined as a response to those fundamental cultural beliefs that underlie this continuing sorcery. Such a conscionable Western confession seems light-years away if one takes the results of the 2008 Canadian election as any indication.

CONCLUSION: CONFESSIONS DENIED

After dissolving Parliament on September 7, 2008, Prime Minister Harper defined the electoral choice Canadians were to face: "They will choose between clear direction or uncertainty, between common sense or risky experiments, between steadiness or recklessness."[103] The reckless risk he referred to was Dion's Green Shift carbon plan, which the Liberal leader proposed would "be good for the environment and good for the economy" by supporting a shift to more energy-efficient practices.[104] In the spirit of Paul Martin's global conscience, Dion planned to assist northern and rural Canadians, like Inuit, who would be especially impacted by the cost of rising fuel, while providing to all economic incentives for dealing with the injustices of climate change by reducing their carbon footprint.[105]

Historically extending Martin's connection of American Republicanism with Harper's Conservatives, he also argued that "once the Bush administration is gone, our biggest trading partner will move towards a greener future" and it would be good if Canada was prepared to be a leader in the evolving international carbon economy.[106] From this Liberal perspective, Prime Minister Harper's epic denial of climate change and promotion of the tar sands was ill-prepared to lead Canada into a new economic era. The past few chapters have similarly suggested that the Conservative government's overly rational focus on conserving Canada's frontier economy in the form of the tar sands and Northwest Passage can be seen as particularly "uncertain, risky and reckless," especially since it requires denying climate research to maintain this direction. While this Conservative denial can be defined as a broken *tirigusuusiit* that is supporting unsustainable behaviours, Dion's progressive Green Shift plan has its own deeply historic inconsistencies that were displayed once the election began.

The Liberal's first green campaign was undermined early on by the public optics of a carbon plan being flown back and forth across the nation in "a 30 year old gas-guzzling Boeing 737 owned by Air Inuit" that was 35 percent less fuel-efficient than the Conservative plane.[107] This regrettable situation was used by Dion as an opportunity to publicly display how his plan proposed to ensure that the "true cost" of pollution is accounted for in a way that "will be good for our economy and the environment."[108] His approach was to purchase carbon offset credits and challenge Prime Minister Harper "to offset his emissions as well"[109]—a challenge that was refused by the Conservative leader. Carbon offsetting is an economic tool whereby people pay to neutralize fuel-inefficient activities like air travel by supporting tree-planting, conserving old-growth forests, renewable energy initiatives, or energy-efficiency activities. It was, as such, an approach completely consistent with the Green Shift's central proposal of utilizing a cap-and-trade system that would make companies and individuals pay taxes for emissions above a certain level and rebate money for those whose emissions fell below.

The environmental community has debated the validity of carbon offsets, with some only supporting those that build society's capacity to reduce fossil fuel use "through renewable energy and energy efficiency."[110] While these were the type of carbon offsets Dion chose, even such a response has its critics. For example, Kevin Smith of Carbon Trade Watch argues that carbon offsets send "the message that you can have your 'carbon cake' and eat it too."[111] Some even define these initiatives as "carbon colonialism" because they ultimately maintain material privilege in a time when emission reductions are required as part of an environmentally just response.[112] Rather than taking Gaia's and *Sedna's* darkening changes as signs of the need to confess broken *tirigusuusiit*, Dion's Liberal option remained faithful to a Western formal economy that the preceding chapters describe as needing a more socially conscionable contextualization.

While Dion had chided Prime Minister Harper during the campaign for not having "faith in Canadians' ability to meet great challenges,"[113] the election result on October 14 suggested he had misread the environmental motivations and capacities of Canadians. As one columnist wrote the day after the election, the winning Conservatives had "barely mentioned the environment, except when it was to disparage the green policies of its opponents."[114] This is despite the fact that, prior to the election call, the economy and climate ranked as the top two concerns of Canadians. Looking at the electoral success of a Conservative climate plan that experts believe cannot meet even its limited goal of reducing emissions by 20 percent of 2006 levels by 2020, editorialist Jeffrey Simpson concluded its appeal is "that it demands nothing of voters."[115] Supporting the view that Dion's faith in Canadians to meet this climate challenge was misplaced, one of his Liberal advisors described the Green Shift carbon tax plan as "a millstone around his neck."[116] With Canada's Liberals increasingly wary of climate platforms and the public seemingly desirous of a sustainable change that will not impact their frontier economy or standard of living, it is highly likely that Canada will, in the foreseeable future, internationally lag in enacting any significant carbon policy or confessions.

If taking responsibility for carbon behaviours in a context of the Liberal promotion of a growth economy is not appealing, then it is even more unlikely that Canadians would willingly elect such a carbon policy option in a no-growth or steady-state economic vision. Yet this book has suggested the depth of today's challenge may require just such an apocalyptic change to the West's wasteland approach. Even if people agree that there is the need to respond in such an economically transformative sense, its political viability will likely become more embattled by the need to respond to justice issues like those raised by the Inuit. Slowing down the economy is one thing, but doing that in the context of redistributing wealth to deal with global inequalities rooted in the colonial, development, and sustainable development past would be a truly epic change in our cultural conscience. In light of this significant challenge, I think it is incumbent upon today's researchers to symbolically follow the Inuit shaman and other spiritual guides into the depths of *Sedna* and Gaia so as to increase our awareness of the cultural and religious blocks to enacting a global conscience. Considering the powerful hold fossil fuels still have on the economizing minds of Canadians, Americans, and Westerners, it does seem that a broad interdisciplinary and intercultural endeavour will be required to inspire such cultural, not just carbon, confessions.

In his farewell speech to Liberals, Dion continued to support the Green Shift carbon response by pointing out that "we still have a climate change problem" and whoever is the new leader is "going to have to deal with that."[117] He was correct in making such an assertion, for, as I have argued throughout this book, without a conscionable response to our role in climate change there may be significant apocalyptic implications for humans and nonhumans the world over. Interestingly, this dark potential may also be our one saving grace as Gaia's indeterminate forces intervene to make limited cultural beliefs and practices accountable. As Kevin Phillips and Jared Diamond displayed in the previous chapter, the interaction of political, economic, cultural, ecological, and climate factors have the potential to initiate apocalyptic changes in a culture. It would be more ideal if we chose to make these cultural changes, but with each passing day it does seem more likely that today's global and regional changes may well force us into confessions not of our choosing. The perplexing question that IQ has spent a long time struggling with, and which Western climate research and policy now needs to come to terms with, concerns how to respond to such an animated and indeterminate existence. With *Sedna*'s regional and Gaia's global hair becoming smoggier with each passing day, it is necessary for us to figure out how we are going to enact both carbon and cultural confessions in relation to global and regional realities.

Before concluding with my final thoughts on the future of climate research and policy in Canada and the world, I want to symbolically move us in that direction by returning once more to Rasmussen and the image of him repeatedly asking a shaman to draw *Sedna*. As the shaman finally undertook the task, Rasmussen wrote that he trembled in fear.[118] Though there are today many images of *Sedna*, Arnakak explained to me that the shaman was at this time "in a state of great distress" because representing a powerful ancestor in this way was seen as dangerous due to its lack of humility.[119] She is after all "both giver and taker of life," the ancestor who teaches Inuit that no individual or community "can claim control over Nature."[120] It was only with the colonial pressures of the twentieth century that Arnakak states, "Inuit started overtly portraying *Sedna* in a mermaid form to preserve some of the tradition."[121]

The value of this representational humility, which was discussed earlier in relation to hunting rituals, is highlighted today in proliferating ecological issues like the ones Rappaport mentions: "species extinctions, atmospheric warming and the social and political disruptions."[122] He writes that since ecological processes "are in some degree unknown and in even larger degree unpredictable," it may be more adaptive "to drape such processes in supernatural veils than to expose them to the misunderstandings that may be encouraged by empirically accurate but incomplete" research and policy.[123] With the IPCC and *Arctic Climate Impact Assessment* continuing to represent animals and climate as "goods and services," it seems much Western climate research and policy is still displaying a lack of humility that is central to *Silatuniq*. This wasteland approach, which bears the signs of unheeded *tirigusuusiit*, will need to be dealt with if we are to envision a conscionable response to the greenhouse gas sorcery that underlies Gaia's climate changes, *Sila*'s northern warming, and *Sedna*'s receding animals.

ENDNOTES

1   CBC News, "Gore: Nobel Prize Win Shows Climate Change
a 'Planetary Emergency,'" 12 October 2007, http://www.cbc.ca/
world/story/2007/10/12/nobel-peace.html.
2   Ibid.
3   *An Inconvenient Truth*, directed by Davis Guggenheim,
Paramount Classics, 2006.
4   CBC News, "Gore: Nobel Prize Win."
5   Quoted in John Bennett and Susan Diana Mary Rowley,
Uqalurait: An Oral History of Nunavut (Montreal: McGill-
Queen's University Press, 2004), 172.
6   Daniel Merkur, *Powers Which We Do Not Know: The Gods
and Spirits of the Inuit* (Moscow, ID: University of Idaho Press,
1991), 102.
7   Susan Joy Hassol et al., *Impacts of a Warming Arctic: Arctic
Climate Impact Assessment (ACIA)* (Cambridge: Cambridge
University Press, 2004), 60; also see O. Anisimov and B.
Fitzharris, "Polar Regions (Arctic and Antarctic)," in contri-
bution of Working Group II to the TAR of the IPCC, *Climate
Change 2001: Impacts, Adaptation, and Vulnerability* (Cam-
bridge: Cambridge University Press, 2001); Habiba Gitay et al.,
"Ecosystems and Their Goods and Services," in contribution of
Working Group II to the TAR of the IPCC, *Climate Change 2001:
Impacts, Adaptation and Vulnerability* (Cambridge: Cambridge
University Press 2001); Erika S. Zavaleta and Jennifer L. Royval,
"Climate Change and the Susceptibility of U.S. Ecosystems to
Biological Invasions: Two Cases of Expected Range Expan-
sion," in S. H. Schneider and T. L. Root, eds., *Wildlife Responses
to Climate Change: North American Case Studies* (Washington,
DC: Island Press, 2002).
8   Quoted in Charles P. Wohlforth, *The Whale and the Supercomputer* (New York: North Point Press, 2004), 244.
9   Gitay et al., "Ecosystems and Their Goods and Services," 25;

also see Zavaleta and Royval, "Climate Change and the Susceptibility of U.S. Ecosystems to Biological Invasions," 281.

10 Liberal Party, "The Green Shift," http://www.thegreenshift.ca/default_e.aspx (accessed 17 August 2008).

11 Ibid.

12 Prime Minister Stephen Harper, "Dion's Carbon Tax Flip-Flop Will Punish Taxpayers," 19 June 2008, http://www.conservative.ca/EN/1091/100824.

13 Much of this story is based upon *Nuliajuk: Mother of the Sea Beasts*, directed by John Houston, Triad Films, Halifax, NS, 2001; David F. Pelly, *Sacred Hunt: A Portrait of the Relationship between Seals and Inuit* (Vancouver: Douglas & McIntyre, 2001).

14 Pelly, *Sacred Hunt*, 11.

15 *Nuliajuk: Mother of the Sea Beasts*.

16 Pelly, *Sacred Hunt*, 11–12.

17 Ibid.

18 *Nuliajuk: Mother of the Sea Beasts*.

19 Pelly, *Sacred Hunt*, 11–12.

20 E-mail correspondence, 14 July 2005.

21 Pelly, *Sacred Hunt*, 12; Merkur, *Powers Which We Do Not Know*.

22 E-mail correspondence, 29 September 2005.

23 Merkur, *Powers Which We Do Not Know*, 255.

24 Mircea Eliade, *Shamanism: Archaic Techniques of Ecstasy* (Bollingen Series, Princeton, NJ: Princeton University Press, 1972), 294.

25 E-mail correspondence, 29 September 2005.

26 E-mail correspondence, 17 March 2004; 30 September 2005.

27 Nurit Bird-David, "Animism Revisited: Personhood, Environment, and Relational Epistemology," *Current Anthropology* 40 (1999); Nurit Bird-David, "Beyond the Original Affluent Society: A Culturalist Reformulation," *Current Anthropology*

33, no. 1 (1992). For other significant analyses on the relation of hunter-gatherer ways of living to animism, see Tim Ingold, "Rethinking the Animate, Re-Animating Thought," *Ethnos* 71, no. 1 (2006); and Tim Ingold, "On the Social Relations of the Hunter-Gatherer Band," in R. B. Lee and R. Daly, eds., *The Cambridge Encyclopedia of Hunters and Gatherers* (Cambridge: Cambridge University Press, 1999).

28 Edward Burnett Tylor, *Primitive Culture* (New York: Harper, 1958).

29 Bird-David, "Animism Revisited," s68.

30 Bird-David, "Animism Revisited," s69; also see Linda Tuhiwai Smith, *Decolonizing Methodologies: Research and Indigenous Peoples* (London: Zed Books, 1999), 99.

31 Bird-David, "Animism Revisited," s78.

32 Ibid., s77-s78.

33 E-mail correspondence, 30 September 2005.

34 Ibid.

35 Ibid., s71.

36 Ibid., s73.

37 Anne Primavesi, *Gaia and Climate Change: A Theology of Gift Events* (London: Routledge/Taylor & Francis Group, 2009), 72.

38 Ibid., 72.

39 Gearóid Ó Crualaoich, *The Book of the Cailleach: Stories of the Wise-Woman Healer* (Cork, Ireland: Cork University Press, 2003), 101–102.

40 Bird-David, "Animism Revisited," s73.

41 Bird-David, "Beyond the Original Affluent Society," 32.

42 Aqqiaruq quoted in S. Fox, "These Are Things that Are Really Happening: Inuit Perspectives on the Evidence and Impacts of Climate Change in Nunavut," in I. Krupnik and D. Jolly, eds., *The Earth Is Faster Now: Indigenous Observations of Arctic Environmental Change* (Fairbanks, AK: Arcus, 2002), 43–44.

43 Ibid.

44 *Impacts of a Warming Arctic: Arctic Climate Impact Assessment*, 74.

45 Gitay et al., "Ecosystems and Their Goods and Services," 277.

46 Terry L. Root and Stephen H. Schneider, "Climate Change: Overview and Implications for Wildlife," in S. H. Schneider and T. L. Root, eds., *Wildlife Responses to Climate Change: North American Case Studies* (Washington, DC: Island Press, 2002), 2–3.

47 Gitay et al., "Ecosystems and Their Goods and Services"; Root and Schneider, "Climate Change: Overview and Implications for Wildlife," 20.

48 *Impacts of a Warming Arctic: Arctic Climate Impact Assessment*, 59.

49 Ibid., 58.

50 Quoted in Merkur, *Powers Which We Do Not Know*, 103.

51 Quoted in Bennett and Rowley, *Uqalurait*, 171–172.

52 *Impacts of a Warming Arctic: Arctic Climate Impact Assessment*, 58.

53 John Robert McNeill, *Something New under the Sun: An Environmental History of the Twentieth-Century World* (New York: W. W. Norton and Company, 2000), 262.

54 Gitay et al., "Ecosystems and Their Goods and Services," 271.

55 McNeill, *Something New under the Sun*, 263.

56 Gitay et al., "Ecosystems and Their Goods and Services," 272.

57 McNeill, *Something New under the Sun*, 263.

58 Ibid.

59 E-mail correspondence, 26 September 2005.

60 Pelly, *Sacred Hunt*, 26.

61 Ibid.

62 E-mail correspondence, 17 February 2004.

63 Quoted in Bennett and Rowley, *Uqalurait*, 43.

64 Ibid.

65 George Wenzel, *Animal Rights, Human Rights: Ecology, Economy and Ideology in the Canadian Arctic* (London: Belhaven Press, 1991), 139.

66 Pelly, *Sacred Hunt*, 106.

67 For example, see Roy A. Rappaport, *Ritual and Religion in the Making of Humanity* (Cambridge: Cambridge University Press, 1999); Bird-David, "Animism Revisited"; Bird-David, "Beyond the Original Affluent Society."

68 Bird-David, "Beyond the Original Affluent Society," 32.

69 George W. Wenzel, "Ninqiqtuq: Inuit Resource Sharing and Generalized Reciprocity in Clyde River, Nunavut," *Arctic Anthropology* 32, no. 2 (1995).

70 Quoted in Bennett and Rowley, *Uqalurait*, 78.

71 Mark Nuttall, "Indigenous Peoples, Self-determination and the Arctic Environment," in M. Nuttall and T. V. Callaghan, eds., *The Arctic: Environment, People, Policy* (Amsterdam: Harwood Academic Publishers, 2000), 392.

72 Quoted in Bennett and Rowley, *Uqalurait*, 45.

73 Hugh Brody, *Maps and Dreams: Indians and the British Columbia Frontier* (Vancouver: Douglas & McIntyre, 1981), 37.

74 Ibid.

75 Ibid.

76 Hugh Brody, *The Other Side of Eden: Hunters, Farmers and the Shaping of the World* (Vancouver: Douglas & McIntyre, 2000), 133.

77 Ibid., 260.

78 Ibid.

79 E-mail correspondence, 6 October 2005.

80 E-mail correspondence, 7 February 2005.

81 Ibid.

82 Nuttall, "Indigenous Peoples, Self-determination and the Arctic Environment," 393.

83 Ibid.

84 Pelly, *Sacred Hunt*, 60.

85 Rappaport, *Ritual and Religion in the Making of Humanity*.

86 Quoted in Joseph Campbell, *Primitive Mythology* (Harmondsworth, UK: Penguin Books, 1976), 294.

87 Rappaport, *Ritual and Religion in the Making of Humanity*, 261.

88 E-mail correspondence, 16 June 2004.

89 Rappaport, *Ritual and Religion in the Making of Humanity*, 268.

90 Merkur, *Powers Which We Do Not Know*, 198.

91 Pelly, *Sacred Hunt*, 26.

92 *Nuliajuk: Mother of the Sea Beasts*.

93 Ibid.

94 E-mail correspondence, 12 July 2005.

95 This paragraph is based on the citation of Knud Rasmussen's research in Eliade, *Shamanism*, 294–296.

96 Merkur, *Powers Which We Do Not Know*, 114.

97 Eliade, *Shamanism*, 294–296.

98 E-mail correspondence, 16 June 2004.

99 *Nuliajuk: Mother of the Sea Beasts*.

100 Rappaport, *Ritual and Religion in the Making of Humanity*, 268.

101 E-mail correspondence, 8 July 2004; 12 July 2004.

102 Brody, *Maps and Dreams*, 37.

103 Karen Howlett, S. Chase, and D. Leblanc, "Battle Begins for Elusive Majority," *Globe and Mail*, 8 September 2008, sec. A, p. 1.

104 Liberal Party, *The Green Shift: Building a Canadian Economy for the Twenty-First Century*, 2008, http://www.cbc.ca/newsatsixns/pdf/liberalgreenplan.pdf.

105 Ibid.

106 Ibid.

107 J. Taber, "Liberals Hit Turbulence as Campaign Takes Off," *Globe and Mail*, 8 September 2008, sec. A, p. 1.

108 Liberal Party, *The Green Shift: Building a Canadian Economy*.

109 Taber, "Liberals Hit Turbulence as Campaign Takes Off."

110 Adriana Barton, "Air Canada's Footprint May Be Greater than Zero," *Globe and Mail*, 30 May 2007, sec. L, p. 2.

111 Heidi Sopinka, "Carbon Offsets: A Shell Game?" *Globe and Mail*, 30 May 2007, sec. L, p. 2.

112 Ibid.

113 M. Campbell et al., "Dion's Green Plan Would 'Wreak Havoc,'" *Globe and Mail*, 12 September 2008, sec. A, p. 1.

114 Gary Mason, "The Environment Was Not a Winning Issue on this Campaign Trail," *Globe and Mail*, 15 October 2008, sec. A, p. 7.

115 Jeffrey Simpson, "When All's Said and Done the Carbon Tax Is Toast," *Globe and Mail*, 22 October 2008, sec. A, p. 21.

116 Jane Taber, "Dion Ignored Advisers' Advice, Preferring to Act as 'a Lone Wolf,'" *Globe and Mail*, 17 October 2008, sec. A, p. 9.

117 Jane Taber, "Abstaining no Longer a Liberal Option," *Globe and Mail*, 24 October 2008, sec. A, p. 4.

118 Cited in *Nuliajuk: Mother of the Sea Beasts*.

119 E-mail correspondence, 9 June 2006.

120 E-mail correspondence, 29 September 2005.

121 E-mail correspondence, 9 June 2006.

122 Rappaport, *Ritual and Religion in the Making of Humanity*, 452.

123 Ibid.

CONCLUSION: OUR CLIMATIC CHALLENGE

On my first evening in Chesterfield Inlet, Simionie Sammurtok referred to some advice his grandmother offered before dying a number of years earlier:

> My grandson, I am going to tell you something that you should remember all the time. When animals come from the tree-line, going north, when this happens, you have to leave it in your heart and in your mind.

In the months that followed, I found myself circling back to these words that were so important to Sammurtok. Unsure of what "leaving it in your heart and mind" could mean, I asked Jaypeetee Arnakak his thoughts. He responded:

> In IQ, it is believed that the physical body is aware of the environment way more than the conscious ego—which one must regard with a scepticism and forbearance. Inuit believe that the body will crave and visualize what it physiologically and spiritually requires to heal the body. I would suspect that what Simionie is saying is that ... the human heart has the capacity to perceive messages that the ego has up to now ignored.[1]

There is, in his view, a kind of intuitive *Silatuniq* that the heart can offer our limited knowledge. Such a complementarity of rational and emotive understanding was evident at an Inuit elder conference when participants defined IQ as "using heart and head together."[2] According to Sammurtok's grandmother, this integrating practice becomes even more vital in times like today when *Sedna*'s animals are moving northward, *Sila* is warming, and Gaia's climate is changing. Considering these trends in a Canadian and global context, I have similarly concluded that it is time for climate research and politics to marry its abundant interdisciplinary knowledge and policy options with the intercultural inspiration of a heartfelt global conscience.

With the northern ice receding and such a global conscience nowhere to be found in Canadian politics, the Conservative government followed its January 2009 Polar Bear Roundtable with a summer northern sovereignty campaign that displaced concern about *Sila*'s warming with frontier optimism about a melting Northwest Passage. As Prime Minister Harper promised to defend the romantic and economic right of Canadians to the North with armed icebreakers, northern paratroopers, and a Baffin Island deep-water port, the new post-Dion approach of Liberal leader Michael Ignatieff began with a double movement that shifted much closer to this conservativism. Though he described the North as "the world's refrigeration system" and called for Canada and other arctic nations to stabilize the climate,[3] Ignatieff's initial concern was with conserving Canada's frontier economy in a progressive way. This is not only reflected in his thoughts on the power Canada can derive from an open Northwest Passage but also in his proposed restructuring of how the land's extensive energy resources are used. His vision includes processing more tar sands oil in Canada, maintaining energy reserves for difficult times, and creating east-west energy corridors to counteract its current southward flow to the United States. In other words, the nation should continue extracting frontier resources and services from the tar sands and Northwest Passage, but the pace must be slowed in order to sustain Canadian energy security and international sovereignty. As with Prime Minister Martin's failed call for a global conscience, Ignatieff deemed his approach consistent with a sustainable climate response that recognizes Canada can no longer "wait for our environmental policy to be determined outside our borders."[4] Instead of being concerned about the influence of President Bush's denial, Liberals would now critique Prime Minister Harper for following the lead of American politics—even President Barack Obama's newly elected Democratic government.

Until the November 2008 election of President Obama and his campaign promise to deal with the impacts of "dirty oil," Dion's Liberal defeat and Prime Minister Harper's economizing faith in the tar sands suggested even a mediocre Canadian climate response would fade into oblivion. The day after the American election this seemed to change as Prime Minister Harper sent a proposal to strike a joint climate change pact that would protect the tar sands from "new U.S. climate-change rules by offering a secure North American energy supply."[5] Talking to CNN in February 2009, the prime minister suggested Canada has always wanted a "regulatory regime" that could reduce greenhouse gas emissions, but was hindered by the policies of the preceding Republican government.[6] At the July 2009 G8 meeting in Italy, he continued this seeming change of thought by suggesting the Conservative government had consistently said it was illogical to promote a climate response without the United States being involved. Following Obama's lead, the Conservatives would now require cars and small trucks to be more fuel-efficient by regulating vehicle carbon emissions, as well as place a cap on greenhouse gas emissions that would require industries to buy and sell emission permits in a carbon market. These actions led Alberta's provincial Conservatives to ask an interesting question in a media-leaked internal memo: "Why is the federal Liberal leader giving more ringing endorsements about the importance of the Alberta oil sands than our prime minister from Calgary?"[7]

In June 2009, on the one-year anniversary of Prime Minister Harper's apology to the survivors of the colonial residential schools, Ignatieff expanded on his Liberal alternative by stating only when all indigenous peoples "share in the goals and dreams of their fellow Canadians" will Canada be a nation built on justice.[8] In his earlier academic writings, Ignatieff proposes that "the sharing must go both ways" so a mutual recognition of different understandings can rebalance our relations and offer a new equilibrium.[9] He also wrote of the need to replace the conflicts of converting difference with a dialogue based on respect for the world's "thousand different traditions."[10] Unfortunately, there appears to be little in Ignatieff's political embrace of a frontier economy and the tar sands that suggests much will culturally change in Canada if he was elected. This political failure reflects, in my mind, cultural issues that go far beyond the double movements of liberal and conservative politics.

While these pages have, as with Ignatieff, highlighted the value of culturally expanding the voices of today's climate dialogue, the Italian G8 meeting also seemed to hint at this need when it was agreed that the G20 would be a more suitable venue for determining common ground before the pivotal United Nations Copenhagen Climate Change meetings in December 2009. Prior to the G8 meeting, Ignatieff had raised to the media an Environment Canada report that indicated Canada's 33.8 percent rise of greenhouse gas emissions above its 1990 levels was higher than any other G8 nation. Findings like these increasingly entered the public light as the Italian meetings began and Canada's international failure was starkly highlighted by a World Wildlife Fund report that graded the G8 nations and found Canada to now be trailing the United States in climate performance. The Conservatives surely deserve blame for this abysmal failure because of their continued denial of climate research, policy, and leadership, but Ignatieff's Liberal approach is also equally troubling because of the disingenuousness of calling for a global conscience without contextualizing its faith in the economic growth of sustainable development. An even bigger issue is the potential influence of powerful political economic assumptions on interdisciplinary climate research, such as when the Bush government reigned in the *Arctic Climate*

Impact Assessment, the tar sands mediates Canada's political engagement of climate research, or a sustainable development agenda influences the Intergovernmental Panel on Climate Change (IPCC).

Our current economizing belief was briefly challenged in October 2008 when the global economic depression hit during the Canadian and American election campaigns. With many late-2008 newspaper editorials proclaiming the end of American capitalism, John Gray from the London School of Economics wrote that the scrambling "going on in Washington marks the passing of only one type of capitalism—the peculiar and highly unstable variety that has existed in the United States over the past 20 years."[11] As with the societal collapse analyses of Kevin Phillips and Jared Diamond found in Chapter 5, Gray further suggested the interaction of war and debt was resulting in an American empire that was waning just as with the Soviet Union a couple decade ago and the British empire a century earlier.[12] The questioning in the United States led to Alan Greenspan, the former Chief of the United States Federal Reserve, being brought before a congressional committee to explain the role of regulators in the uncertainty. Particularly relevant was his response that "I made a mistake in presuming that the self-interests of organizations, specifically banks and others, were such that they were best capable of protecting their own shareholders."[13] Shocked by these events, he added that he still does "not fully understand why it happened."[14] If he considered the theological critiques of Primavesi or, even more unlikely, the IQ of Arnakak, then the idea that a belief system based on greed can bring such destruction may not have seemed so shocking. For their analyses highlight the social and ecological importance of placing economizing solutions in their proper place as one aspect of climate research and politics. Until such cultural predispositions are dealt with, it is highly likely that our international and national responses will be uninspired and ineffective for the various reasons I have discussed over the preceding chapters.

While there were calls for a more comprehensive response to the 2008 crash, the arising economic scarcity had some predictably constraining results on the 2009 climate dialogues that began in Canada with the January Polar Bear Roundtable and heightened as the international Copenhagen meetings of December approached. Over the preceding month of November, one could hear on Canadian radio the Friends of Science reinvigorate their campaign by pointing out the irrationality of making significant changes in a time of economic uncertainty, especially when recent evidence suggests there has been no warming since 1998. This "no warming" story arose from a BBC report that indicated from 1998 to 2009 there was no evidence of a global warming trend. In Canada, the editorialist Rex Murphy of the CBC and the national *Globe and Mail* newspaper wrote that the "BBC is not the only voice showing sprigs and shoots of independent thinking on global warming," and then offered Ross McKitrick—someone we saw in Chapter 2 as being a close associate of the Friends of Science—as another "mind outside the herd."[15] Critiquing this specific "no warming" trend, environmental thinker David Suzuki pointed out that the time period was cherry-picked by "starting with the warm 1998 and ending with the cold 2008."[16] Other influential environmental thinkers like George Monbiot and Thomas Homer-Dixon similarly argued that choosing "this year as the starting point for a trend is misleading at best and dishonest at worst."[17] Despite sceptically choosing this cooler period, Suzuki added there is still evidence for "a warming trend of 0.11°C per decade."

This pre-Copenhagen debate clearly echoes the analyses of Chapters 2 and 3, which concluded that the difficulty with the research of McKitrick and the Friends of Science is that it is based on a long out-dated approach to scientific uncertainty. There is an indeterminate complexity to Gaian climate change that transcends a scientific law-based approach to planetary economic management. Even if there was evidence for a slow down or reversal in global warming, we saw in Chapter 5 that there is also a need to remember the possibility of Gaia's non-linear discontinuities, such as a shift in the Atlantic Ocean's thermohaline circulation. A better indicator of climate change than a simple linear rise in temperature may be the frequency of broken records reflected in regional and global phenomena like long periods of drought, unexpected temperature highs and lows, and receding northern ice. As climate physicist Spencer Weart pointed out earlier, "when you push on something steadily it may remain in place for a while, then move with a jerk."[18] In other words, we are neither talking about a smooth linear warming trend from one year to the next nor changes that will be similar across all regions. The indeterminism of Gaia and *Sedna* is hard for the modern mentality of many climate deniers, media pundits, politicians, and researchers to appreciate, and as such it is not surprising that many of us find it equally difficult to grasp the importance of the related complementarity etiquette that may be the basis for a sustainable future.

The November before Copenhagen also witnessed another powerful story that increased the cultural uncertainty surrounding climate research. A number of stolen e-mails from the Climatic Research Unit at the University of East Anglia in the United Kingdom indicated its significant contributions to the IPCC may have been, as the Friends of Science argue, compromised by its politicization. "Climategate," as it became known, provided some evidence that scientists were responding to the pressure of the climate change denial industry with a secretive and obscuring approach to sharing problematic data. While sceptics saw this as more evidence for a "no warming" view of things, it also stirred up much uncertainty amongst environmental thinkers. Both Suzuki and Monbiot described these e-mails as being blown out of proportion by the denial industry, and that in the end they only raised questions "about one or two lines of evidence out of several hundred lines of evidence which show that man made climate change is taking place."[19] That said, the practice revealed to Monbiot the need for climate researchers to publicly release their "data so that other people can analyze it." From Anglia's Climate Research Unit, one of the nonimplicated climate researchers, Professor Hulme, similarly expressed concern that "the activist, ideological way that research has been used by the IPCC, has put scientists in the position of being the authors of policy—a position that distorts the role of science in society."[20] This view was echoed in the Canadian media by Rex Murphy who argued it is necessary "that advocacy follows the science, not that science seeks to comport with advocacy."[21] As with Monbiot, Hulme proposed that a response to this untenable situation will require making raw climate data "available to everyone, including climate-change sceptics."[22] More transparency so that climate research can be less ideological and properly contextualized in a broader dialogue is a position held in common by Hulme, Monbiot, the Friends of Science, and the intercultural assessment of this book—though, as we have seen, there are clearly different interpretations of how we define a "broader" dialogue.

Back in 2001, David Demeritt offered a penetrating social analysis of climate research that in many ways foretold the difficulties of Climategate and offers us some insights on a potential road forward. His concern was that interdisciplinary climate research may be undermined by the way it responds to sceptics who are largely "paid by the fossil fuel industry."[23] The effort to win political trust by focusing on scientific facts and certainty may in the end increase public uncertainty because it will, in his words, "invite political opponents to conduct politics by waging war on the underlying science (and scientists!)."[24] This is exactly what Monbiot and Suzuki are documenting when they refer to the way in which the stolen e-mails have been blown out of proportion by the denial industry. Not only does this attack bear out Demeritt's point but the actual questionable e-mails reflect his bigger concern that climate research can become more defensive rather than reflexive in acknowledging "the irreducibly social dimension of scientific knowledge and practice."[25] While sceptics promote denying climate research by exposing its social assumptions and scientific uncertainties, he has found proponents of the IPCC often take the equally extreme position of denying any social influence. Almost foretelling the difficulties of Climategate, Demeritt concludes that the attempt to alleviate public doubt by increasing the population's climate knowledge needs to be replaced with initiatives that increase "understanding of and therefore trust in the social process through which those facts are scientifically determined."[26] The climate's indeterminate nature will always result in regional exceptions and seemingly contradictory findings that sceptics will use as fodder for political inaction, and as such there is a need to figure out a more open response. If there is to be a silver lining to Climategate, it may well depend upon the extent to which climate research takes up such a challenge by more broadly considering a social response to the cultural complexity of our present situation.

It was amidst the rising public uncertainty of Climategate and a "no warming story" that environmentalists, bureaucrats, researchers, cultural advocates, and national leaders descended upon Copenhagen in December with high expectations for a more inclusive climate agreement to replace the embattled Kyoto Protocol. Canadian Prime Minister Harper confirmed his attendance after it became clear American President Obama, Chinese Premier Wen, and many other national leaders would be present. With the negotiations faltering and only a day left, the American President made a plea in his speech to the 120 heads of state: "While the reality of climate change is not in doubt, our ability to take collective action hangs in the balance."[27] This was followed by last-minute negotiations between the United States, China, India, and South Africa that led to the Copenhagen Accord. The small and selective group of nations involved in this negotiation subsequently led critics to raise questions about the waning power of a United Nations process based on international plurality. It was clearly a political approach that moves in the opposite direction from the intercultural inclusiveness Demeritt argues for and which I propose here.

Unlike the Kyoto Protocol that still remains in place for the short term, the Copenhagen accord is a nonbinding political framework that was initially signed by Canada and a total of twenty-eight nations. Its central mission is defined as cooperatively combating climate change to keep the average global temperature increase below 2°C, though there is no deadline for making the agreement legally binding and the reduction commitments of developed countries still need to be negotiated. As with the Kyoto Protocol, issues of equity and responsibility were major obstacles in the negotiating process. Though it was agreed that developed nations would provide billions of audited dollars to help poorer nations implement mitigation and adaptation measures, the source of the money was unclear and it would not arrive until around 2020—all of which further suggested more potential for broken promises in the spirit of Kyoto. This irresponsibility was particularly epitomized by the Canadian government's approach.

During the Copenhagen meetings, Canada's Environment Minister Jim Prentice re-announced the Conservative government's half-hearted proposal of a 20 percent reduction in greenhouse gases below 2006 levels by 2020. Even more significant, Canada would give special treatment to the tar sands so as not to hamper its economic development. Such proposals predictably drew the ire of environmentalists who gave the Canadian government various mock awards for climate inaction. Prior to the conference, the Canadian government was criticized by the likes of Monbiot and Suzuki for becoming overly dependent on a fossil fuel resource that was undermining its international reputation. Beyond a United Nations climate change performance index of the world's sixty richest nations, which found only Saudi Arabia to be performing worse than Canada, Monbiot was also concerned about the extensive pressure Canada was exerting to undermine an international agreement. It is a position reminiscent of Richard Black's assertion in Chapter 3 that in a history of international climate change responses, Canada will figure prominently, first for liberally challenging President Bush's leadership and then for leading that unsustainable conservativism. These meagre Copenhagen promises will ensure that Canada will not meet its Kyoto commitments until well past 2020, while creating a drag on international cooperation. It is not only an ideal delay for those aligned with the fossil fuel industry but is also unjust for those in more compromised positions.

In Copenhagen, the Inuit Circumpolar Council provided its own seemingly conflicted assessment of what a global and northern climate change response will entail. During an interview with the CBC, Jimmy Stotts, the chair of the Inuit Circumpolar Council, argued Inuit-owned fossil fuel and mining projects should be exempted from a global agreement so as to continue fostering the development of their communities. As he stated, it "doesn't seem right for Inuit, who have gotten themselves to this point where they can develop and make better communities, without having access to the money that they make from these industries."[28] It was later clarified by the vice-chair, Aqqaluk Lynge, that this is not the official position of the Inuit Circumpolar Council, for it continues to advocate emission reductions because Inuit, as he stated, "are the ones who have been hit hardest by climate change."[29] Based on this position, the Inuit Circumpolar Council advocated for emission cuts that would keep the global temperature rise below 2°C. There was apparently a misunderstanding in the interview as Stotts later clarified his concern about the high price Inuit continue to pay for both the impacts of climate changes arising from unsustainable behaviour to the south and policy decisions that sidestep the arising justice issues. That said, this misunderstanding also clearly highlights the fact that there are Inuit interests that support a northern industrialism that is in line with

Harper's tar sands and Northwest Passage initiatives—as long as it brings benefits to Inuit.

An intercultural justice perspective more in line with my book's central concern was clarified at Copenhagen by the most visible spokesperson for Inuit interests, both at these meetings and over the past decade of climate negotiations, the former chair of the Inuit Circumpolar Council and Nobel Prize runner-up Sheila Watt-Cloutier. In her words: "As we call on the world to change its ecologically degrading practices, we must not accept those practices at home no matter how desperate our need for jobs or economic development."[30] The social complexity of this moral position was further highlighted by Stotts when he warned Inuit about aligning with environmental groups whose research can undermine their traditional hunt of seals, whales, and polar bears.[31] These moral considerations are consistent with the Inuit Circumpolar Council's earlier critique of the *Arctic Climate Impact Assessment*, the human rights petition against America's Republican government, and the challenge of climate research at Canada's Polar Bear Roundtable. It is evident that many blocks need to be overcome before an intercultural approach to climate-human relations can provide an alternative to the political inaction represented at Copenhagen.

While Watt-Cloutier advocates for a strong climate response that does not replicate implicated cultural behaviours, even the limited Canadian response represented at Copenhagen was portrayed by Prime Minister Harper as being uncertain because of its dependence on the now Democrat-led United States. As he stated after signing the agreement: "If the Americans don't act, it will severely limit our ability to act. But if the Americans do act, it is essential that we act in concert with them."[32] In contrast to this conservative Western morality, Liberal leader Ignatieff opined that Canada needs to be less dependent on America in creating "an aggressive, made in Canada climate-change plan."[33] The voice of Monbiot was more critical of Canada's conservativism when he stated, "I believed that the nation that has done most to sabotage a new climate change agreement was the United States. I was wrong. The real villain is Canada."[34] Though he further clarifies that other nations like China are also a political economic drag on the international climate response, Monbiot's critique still highlights the point that Canadians are making significant moral decisions that will impact their future relations with the international community and Inuit. There were clearly extensive cultural uncertainties being displayed in Copenhagen that go far beyond the limits of a rationalizing mind into the unobjective and, I would argue, maladaptive regions of cultural immorality and religious faith.

Before leaving Copenhagen, President Obama announced an optimistic view of the accord as "the first time in history, all the major economies have come together to take action" on climate change.[35] While the Inuit Circumpolar Council proposed to use the promised climate change adaptation fund, environmentalists concluded that the agreement was largely disappointing and vacuous, though some suggested it was better than nothing. On the other side, Canadian climate sceptics like media pundit Rex Murphy asked, "how a conference that essentially aimed at reordering the world energy economy gave such scant attention to the reasonable doubts about the scientific process opened up by Climategate."[36] Other sceptics dwelled on the real hypocrisy of shipping into Copenhagen an estimated 1,200 limousines from Germany and Sweden for dignitaries, the 140 private jets parked at the Copenhagen airport, and the massive use of air travel that marked a negotiation concerned with sustainable greenhouse gas emissions.

Thinking about all these contradictions, debates, delayed responses, fabrications, secrets, and denials, I am brought back to the analyses of Diamond and Tainter on the complex nature of social collapse. If climate change even approaches some of the more abrupt projections, then perhaps we are witnessing in Kyoto, Montreal, Copenhagen, and beyond some of the key intercultural patterns that are revelatory of a global decline in problem-solving abilities. Rather than choosing to bring an apocalyptic *Silatuniq* upon ourselves in response to Gaia's indeterminate changes, we may very well be witnessing the latest society—of global scale and encompassing various cultures—to let the surrounding world mediate these changes. It is unlikely such a choice will end well for future generations or, as is becoming increasingly clear, ourselves.

A more heart-inspired model of the meetings required today is symbolized for me in the Earth Charter deliberations and their concern with defining a global conscience. Following the 1992 Rio Earth Summit, various United Nations and nongovernmental organizations launched a comprehensive research initiative that drew upon "international law, science, religion, and ethics."[37] The charter's foundational value of participatory collaboration was based on three interrelated factors that have been highlighted in the preceding analysis. First, today's world is "characterized by rapid change, increasing globalization, and growing interdependence" that have ecological, political economic, cultural, and spiritual dimensions.[38] Second, environmental issues like climate change "threaten the foundations of world security."[39] Finally, the indeterminate nature of these multi-scalar problems requires global cooperation in defining a shared set of ethics. As with this book's critiques, it concludes that economic rationality must be contextualized in a broad dialogue concerned with optimizing ecologically sustainable ways of living. Considering the cultural challenge this view presents to existing political economic powers, it is of no surprise that the 2002 presentation of the Earth Charter to the United Nations for international endorsement was declined. Despite this political failure, it still offers climate research a model of what is needed to be "the most open and participatory process ever to have occurred in connection with the preparation of an international document."[40] If we can figure out a way to embrace the kind of global conscience epitomized by the Earth Charter process, then the intercultural expansion of voices could increase our access to an array of wise regional and global responses.

From *Sila*'s northern warming to *Sedna*'s animal changes, we have seen that other kinds of cultural knowledge can expand the current climate research endeavour far beyond a traditional ecological knowledge focus on regional climate change. While the *Silatuniq* and IQ of Inuit provide a deeper sense of how to live in the North, I have also argued that it is their contrast with some unsustainable Western beliefs that makes them equally important in a context of devising responses to Gaia's global changes. This is not to say that the climate research of the IPCC and *Arctic Climate Impact Assessment* is to be denied, as with Canada's and America's contemporary conservative scepticism. Rather, I agree with Demeritt's call to demystify scientific knowledge by highlighting its social construction, while not denying its detailed understanding of very real and troubling climate changes. There is a need to raise important questions about how we understand and live with this complex reality in a more inclusive way, and interdisciplinary climate research can be of significant help in that process if it is understood to be simply one powerful globalizing knowledge system that is in need of cultural and regional contextualization. As one Earth Charter principle puts it, all cultures "have much to learn from the ongoing collaborative search for truth and wisdom."[41] In the 2003 *State of the World* report, Gary Gardner made this exact point by suggesting a global response needs to engage the diverse religious traditions that directly inform 83 percent of the world's population.[42] Beyond the grassroots potential of religious worldviews, he adds that anthropological research indicates in cases where cultures have managed their ecology sustainably "the credit often goes to religious or ritual representation of resource management."[43] Regardless of what religion we are talking about, there co-exists today a diversity of spirited approaches that have the potential to be of adaptive value for people as they contemplate regional and global responses to indeterminate climate changes.

Such a broadened dialogue is doubly important today because of the colonial legacy that was discussed in Chapter 4, and its relation to environmental issues like climate change. We have seen in this Inuit case study that the cultural impacts of colonialism and the more recent global development agenda has resulted in the co-existence of various worldviews in any region of today's world. More than that, there are quite clearly different depths of understanding due not only to the cultural degradation of colonial forces but also differences in practice. As the first chapter clarified, the Inuit shaman traditionally had access to a *Silatuniq* that was not part of everyone's experience. Based on these realities, it can be said that a twenty-first-century Inuit youth needs the educational opportunity to contextualize Western knowledge and IQ within their own inspired cultural stories—whether that be a Christian convert's view on an apocalyptic threat, a traditional *Sedna* understanding of shifting polar bear behaviour, or a shamanic *Silatuniq* of northern warming. Spirited sensibilities and stories, when brought into intercultural dialogue, have the pedagogical potential to ground the globalizing tendencies of Western climate research and politics in a diversity of regionally accessible cultural views, practices, and passions.

Beyond the capacity of other cultural traditions to deepen our sense of a global conscience, I have also argued that these understandings are vital because of the insight they can provide on maladaptive beliefs and practices—from Canada's wasteland approach toward the tar sands to the West's continued promotion of a growth economy. In December 2008, one year prior to Copenhagen, this power of maladaptive religious beliefs to fuel climate change was explicitly displayed on the front page of the *New York Times*. As Prime Minister Harper began his second term and President Obama transitioned America to a more liberal approach, a full-colour photo showed an altar of a Pentecostal Christian congregation in Detroit, Michigan. Upon it was a choir, a young girl in ritual dance, and two large white American-made sports utility vehicles (suvs). With the economic crisis hitting America's car culture hard, the paper's caption read, "praying to save the auto industry." This symbolic use of a tar-sands-guzzling suv not only denies the reality of today's climate crisis but also reinforces a deeply flawed religious belief. The automotive industry historically advertised the suv as a purchase that could provide individuals with the means to get off the industrialized road and attain a spiritual experience in nature.[44] Until the recent rise in gas prices and subsequent economic crisis, this romantic image proved quite successful in drawing Western consumers toward a vehicle that is incompatible with sustaining Gaian ecology and climate. In response to such maladaptive American rituals, President Obama campaigned to rebuild America's car culture and fossil fuel economy around clean, renewable energy sources as a strategy to be less dependent on Middle Eastern and dirty tar sand oil.[45] It was a promise whose fulfillment became more uncertain after China's billion dollar tar sands investments raised familiar security issues in American political circles—a concern that probably had some impact on the limited Copenhagen agreement.

The *New York Times* photo symbolizes for me one of the central points of this book: cultural change will have to go far beyond a political shift in energy sources if we wish to inspire a global, national, and regional conscience. There are clearly some powerful cultural and religious beliefs fuelling contemporary political economic behaviours, and it is necessary to consider the ways in which they continue to nurture ecologically maladaptive ways of living in Canada, America, China, and elsewhere around the planet. In much the same spirit as Chapter 4's proposal that our proximity to powerful fossil fuels is influencing our thought, the Christian car ritual and Canadian frontier faith are symbolic of powerful realities that have significant influence on our contemporary climate research and politics. This is what Anne Primavesi intimates when she describes the way in which all planetary beings, including humanity, are inheriting the input of fossil fuels on "the global climate, just as future generations will inherit and be influenced by our inputs into it."[46] Drawing support from liberation theology, she advocates changing our sinful impact on Gaia and its vulnerable populations by contemplating "the complex conditions under which commodities are produced, the ways in which money is acquired and the means by which capital is expanded."[47] She is not alone in suggesting religion can increase our culture's adaptive capacity, for both the Earth Charter and Gardner conclude that common intercultural teachings on the religious value of living simply can be combined with more recent ecological research to "situate consumption in a comprehensive understanding of what it means to be a developed society."[48] Insights like these provide us with a positive sense of what the spirited passion reflected in the Christian SUV ritual or the Inuit tradition of confessing to *Sedna* could potentially add to our already prolific interdisciplinary climate knowledge. There is within our grasp a diversity of rituals and stories that could accessibly raise awareness about the relation of Gaia, *Sila*, fossil fuels, greenhouse gases, and our vehicle purchases or gas-pumping.

In the summer prior to the Copenhagen meetings, America's Democratic government provided an updated scientific view on what Gaian climate change may bring forth in the dark hues of "[r]ising sea levels, sweltering temperatures, deeper droughts, and heavier downpours."[49] Though the administration was optimistic about the possibility of preventing some of the worst consequences, this position was based on softening the original report's finding that "tipping points have already been reached and have led to large changes."[50] The start of these potentially apocalyptic changes in Canada's North was sensed well over a decade ago by Inuit like Sammurtok's grandmother, who told him that "Christ is coming soon because the world is changing." In such times as these, he was advised to infuse knowledge with a heartfelt approach that can sense what is happening in a quickly changing world. This approach is not totally foreign to the West, for we saw that a concern with Gaian climate change also led Primavesi to bring together science, political economy, and theology to ask compassionate questions about the inheritance humanity is passing on to present inhabitants of vulnerable planetary regions and future generations of the whole biosphere—including humanity. Such an inspired endeavour is also intimated in an Earth Charter process that I think models for us a broader interdisciplinary and intercultural way of responding to regional and global climate changes.

We could continue examining humanity's evolving climate change response by considering the shift from a failed Kyoto to an uninspiring Copenhagen, the response of Canadians to Harper's epic failure in leadership, the implications of President Obama's emerging climate policy on the Canadian tar sands and the Copenhagen accord, the debates among Inuit about an effective climate response, or the scientific response to a public Climategate. This climatic story is clearly still emerging, but my book's basic point about the intercultural challenge of climate change has been made. Just as with Bush, Gore, Harper, and Martin, the next ten to fifty years of changing politics will tell us much about the West's, America's, and Canada's cultural relation to the climate. Innovative greenhouse gas reduction policies and technical solutions—such as carbon storage or painting roads white to reflect the sun's energy as the receding northern ice once did—may respectively reduce fossil fuel use and be a long-term stabilizing influence on the climate, but without morally responding to our implicated cultural beliefs we will probably just continue extending today's sorcery. Expanding roadways in a carbon-reduced era will still undermine Gaia's and *Sedna's* indwelling biodiversity, and there is nothing to say that President Obama's meagre liberalizing influence will not be replaced in two, six, or more years with a resurgent conservative rush to the pumps in America, Canada, and other ecologically backwards nations. Rather than moving toward such a culturally insular and, I would suggest, antisocial conservativism, this journey that began at the Polar Bear Roundtable has led me to conclude that our future climate meetings will need to go far beyond Copenhagen if we are to enact a global conscience that can transcend a Western and, even, human frame of reference.

To end this book, I want to consider one last Inuit story on human-polar bear relations that brings us full circle to where we began, but with a potentially broader cosmological sense of the cultural changes needed in Western nations like Canada today. In this story, a bear spirit that was killed by an Inuit hunter has come back to warn a young bear about the power of *kanaaqiajuit* or humans. As we saw in the previous chapter and in Wenzel's introductory polar bear hunt, when animals are hunted properly and ritually propitiated, Inuit thought they would return to their own kind.[51] With a face scarred from his death and consumption, the ancestral bear wisely counsels the younger one that human powers should not be thought of in terms of our canine teeth that require us to kill in close proximity. In contrast, human weapons are "much longer than their height."[52] While the bear was referring to the harpoons and guns of the Inuit, many of today's bear spirits that smog *Sedna*'s hair are, according to many climate researchers and environmentalists, being struck down by greenhouse gases emitted by even farther removed Canadian, American, Western, and international behaviours.

Reinterpreting such a polar bear story in relation to Inuit and Western understandings of our present situation is valuable because it has the potential to remind us that surrounding our human-focused climate meetings is a whole cosmology of beings talking about us. Building upon Demeritt's analysis, we need to realize it is not merely researchers, sceptics, politicians and, more broadly, humans who are socially constructing a complex reality but that our research and debates are arising in response to the even more encompassing social dialogue of ecological and climatic realities. Rather than humans and, more specifically, Westerners being the absolute and most important holders of knowledge, this book has led me to conclude that there is a need for our political, scientific, and cultural leaders to recognize their physical participation in a dialogue with Gaian climate change, fossil fuels, *Sila*, *Sedna*, other regional indwellers, and ecological beings like the polar bear.

Either by conscionable choice or apocalyptic force, the maladaptive beliefs of Canadians, Westerners, and, more generally, humans are being called to a change of mind that, in James Lovelock's words, can "wonder about the Earth and the life it bears, and speculate about the consequences of our own presence here."[53] The Inuit voices in these pages have shown us that such speculation is fundamental to a *Silatuniq* that can first attend and then respond to *Sila* and, as I have argued, Gaia's climatically responsive cosmology. Climate research and politics can either spiritually broaden the intercultural and interdisciplinary voices involved in our future meetings so as to attend Gaia's global and regional indeterminism, or continue with an overly rational and decontextualized approach that will make the coming apocalyptic change increasingly not of our choosing. We are clearly in need of a heartfelt response not unlike the advice of Sammurtok's grandmother that it is time "to leave it in your heart and mind," or of Jaypeetee Arnakak's description of IQ as a knowledge that "venerates nature as the source and end of life."[54] Promoting such a wise global, national, and regional change in our cultural conscience and consciousness is today's climatic challenge.

ENDNOTES

1   E-mail correspondence, 24 March 2005.

2   Quoted in Fikret Berkes, *Sacred Ecology: Traditional Ecological Knowledge and Resource Management* (Philadelphia: Taylor & Francis, 1999), 6.

3   Michael Ignatieff, *True Patriot Love: Four Generations in Search of Canada* (Toronto: Penguin Canada, 2009), 170.

4   Michael Ignatieff, "Statement from Liberal Leader Michael Ignatieff on International Earth Day," 22 April 2009, http://www.liberal.ca/en/newsroom/media-releases/15747_statement-from-liberal-leader-michael-ignatieff-on-international-earth-day.

5   Shawn McCarthy and C. Clark, "Ottawa Swoops in with Climate-Change Offer," *Globe and Mail*, 6 November 2008, sec. A, p. 1; also see Jeffrey Simpson, "Little New for Obama in Ottawa's Energy 'Offer,'" *Globe and Mail*, 12 November 2008, sec. A, p. 23.

6   Campbell Clark and Brian Laghi, "Harper Encouraged by Obama's Climate Policy," *Globe and Mail*, 19 February 2009, sec. A, p. 13.

7   Jason Fekete and Don Braid, "Stelmach Tories Praise Ignatieff over Harper," *Calgary Herald*, 9 June 2009.

8   Michael Ignatieff, "Statement from the Liberal Leader Michael Ignatieff on the First Anniversary of the Residential Schools Apology," 11 June 2009, http://www.liberal.ca/en/newsroom/media-releases/15921_statement-from-liberal-leader-michael-ignatieff-on-the-first-anniversary-of-the-residential-schools-apology.

9   Michael Ignatieff, *The Rights Revolution* (Toronto: Anansi, 2000), 84.

10   Ibid., 141.

11   John Gray, "America's Global Fall from Grace," *Globe and Mail*, 1 October 2008, sec. A, p. 21.

12   Ibid.

13 Barrie McKenna, "Greenspan Admits 'Mistake' on Bank Regulation," *Globe and Mail*, 24 October 2008, sec. A, p. 1.

14 Ibid.

15 Rex Murphy, "Be Brave: Escape the Climate Box," *Globe and Mail*, 17 October 2009, sec. A, p. 27; also see Margaret Wente, "Why People Are Chilled by Warming," *Globe and Mail*, 15 October 2009, sec. A, p. 21.

16 David Suzuki with Faisal Moola, "Canada Must Do More to Confront Climate Crisis," *Canoe Network*, 20 November 2009, http://cnews.canoe.ca.

17 Thomas Homer-Dixon and Andrew Weaver, "Responding to the Skeptics," *Globe and Mail*, 7 December 2009, sec. A, p. 15; George Monbiot, "Canada's Image Lies in Tattters: It Is Now to Climate what Japan Is to Whaling," *The Guardian*, 30 November 2009, http://guardian.co.uk.

18 Spencer Weart, *The Discovery of Global Warming* (Cambridge: Harvard University Press, 2003), 138.

19 Monbiot, "Canada's Image Lies in Tattters"; also see Suzuki with Moola, "Canada Must Do More to Confront Climate Crisis."

20 Cited in Doug Saunders, "Copenhagen Summit: Breach in the Global-Warming Bunker Rattles Climate Science at Worse Time," *Globe and Mail*, 5 December 2009, sec. A, p. 22.

21 Rex Murphy, "Through Copenhagen's Looking Glass," *Globe and Mail*, 19 December 2009, sec. A, p. 25.

22 Cited in Saunders, "Copenhagen Summit," A22.

23 David Demeritt, "The Construction of Global Warming and the Politics of Science," *Annals of the Association of American Geographers* 91, no. 2 (2001): 328.

24 Ibid.

25 Ibid., 309.

26 Ibid., 329.

27 Cited in Eric Reguly and Shawn McCarthy, "The Copenhagen Summit: Agreement on Climate Reached—but Unfinished," *Globe and Mail*, 19 December 2009, sec. A, p. 22.

28 Jane George, "Inuit Leaders at Odds over Oil and Gas Emissions: Greenland Wants More Development of Carbon-spewing Industries," *Nunatsiaq News*, 13 December 2009, http://www.nunatsiaqonline.ca.

29 Ibid.

30 Cited in ibid.

31 Jane George, "ICC Moves to Patch Up Inuit Climate Change Rift: There Is a Paradox of Development among Inuit," *Nunatsiaq News*, 17 December 2009, http://www.nunatsiaqonline.ca.

32 Cited in Shawn McCarthy, "Canada's Strategy: Promise Now, Implement Later," *Globe and Mail*, 19 December 2009, sec. A, p. 18.

33 Cited in McCarthy, "Canada's Strategy," A18.

34 Monbiot, "Canada's Image Lies in Tatters."

35 Cited in Reguly and McCarthy, "The Copenhagen Summit," A22.

36 Murphy, "Through Copenhagen's Looking Glass," A25.

37 Steven C. Rockefeller, "Global Interdependence, the Earth Charter, and Christian Faith," in D. Hessel and L. Rasmussen, eds., *Earth Habitat: Eco-Injustice and the Church's Response* (Minneapolis: Fortress Press, 2001), 105, 109.

38 Ibid., 102.

39 Ibid.

40 Ibid., 107.

41 Earth Charter in appendix to Rockefeller, "Global Interdependence, the Earth Charter, and Christian Faith," 215.

42 Gary Gardner, "Engaging Religion in the Quest for a Sustainable World," in L. Starke, ed., *State of the World 2003* (New York: W. W. Norton and Company, 2003).

43 Ibid., 156.

44 Richard K. Olsen, Jr., "Living Above It All: The Liminal Fantasy of Sport Utility Vehicle Advertisements," in M. Meister and P. M. Japp, eds., *Enviropop: Studies in Environmental Rhetoric and Popular Culture* (Westport: Praeger, 2002), 181, 188.

45 Gary Mason, "The Environment Was Not a Winning Issue

on this Campaign Trail," *Globe and Mail*, 15 October 2008, sec. A, p. 7; S. McCarthy and C. Clark, "Ottawa Swoops in with Climate-Change Offer," *Globe and Mail*, 6 November 2008, sec. A, pp. 1, 8.

46 Anne Primavesi, *Gaia and Climate Change: A Theology of Gift Events* (London: Routledge/Taylor & Francis Group, 2009), 68.

47 Anne Primavesi, *Sacred Gaia: Holistic Theology and Earth System Science* (London: Routledge, 2000), 100.

48 Gardner, "Engaging Religion in the Quest for a Sustainable World," 167-168.

49 Seth Borenstein, "White House Report on Climate Change Has a Warning, and a Silver Lining," *Globe and Mail*, 17 June 2009, sec. A, p. 15.

50 Ibid.

51 Peter Anatsiaq, quoted in John Bennett and Susan Diana Mary Rowley, *Uqalurait: An Oral History of Nunavut* (Montreal: McGill-Queen's University Press, 2004), 44.

52 Ibid.

53 James E. Lovelock, *Gaia, a New Look at Life on Earth* (Oxford: Oxford University Press, 1979), 12.

54 E-mail correspondence, 1 June 2004.

Abram, David. *The Spell of the Sensuous: Perception and Language in a More-than-Human World*. New York: Pantheon Books, 1996.

Amarok's Song: Journey to Nunavut. Directed by Martin Kreelak and Ole Gjerstad. Words and Pictures Video, in association with the National Filmboard of Canada and the Inuit Broadcasting Corporation, 1998.

Ambrose, Rona. "Clean Air for All Canadians." 19 October 2006. http://www.conservative.ca/EN/2459/56332.

An Inconvenient Truth. Directed by Davis Guggenheim. Paramount Classics, 2006.

Anisimov, O., and B. Fitzharris. "Polar Regions (Arctic and Antarctic)." In *Climate Change 2001: Impacts, Adaptation, and Vulnerability*, contribution of Working Group II to the TAR of the IPCC, 800–841. Cambridge: Cambridge University Press, 2001.

Appleby, Joyce Oldham. *Economic Thought and Ideology in Seventeenth Century England*. Princeton, NJ: Princeton University Press, 1978.

Arnakak, Jaypeetee. *A Case for Inuit Qaujimanituqangit as a Philosophical Discourse*. Iqaluit, NU: JPT Consulting, 2004.

Austin, Andrew, and Laurel Phoenix. "The Neoconservative Assault on the Earth: The Environmental Imperialism of the Bush Administration." *Capitalism Nature Socialism* 16, no. 2 (2005): 25–44.

Baede, A. P. M., et al. "The Climate System: An Overview." In *Climate Change 2001: The Scientific Basis*, contribution of Working Group I to the Third Assessment Report of the Intergovernmental Panel on Climate Change, 85–98. Cambridge: Cambridge University Press, 2001.

Barbour, Ian G. "Scientific and Religious Perspectives on Sustainability." In *Christianity and Ecology: Seeking the Well-being of Earth and Humans*, edited by D. T. Hessel and R. R. Ruether, 385–401. Cambridge: Harvard University Press, 2000.

———. *Issues in Science and Religion*. 1st Torchbook ed. New York: Harper and Row, 1971.

Barton, Adriana. "Air Canada's Footprint May Be Greater than Zero." *Globe and Mail*, 30 May 2007, sec. L, p. 2.

Baum, Gregory. *Karl Polanyi on Ethics and Economics*. Montreal: McGill-Queen's University Press, 1996.

Behringer, Wolfgang. *Witches and Witch-Hunts: A Global History*. Themes in History. Cambridge, UK: Polity Press, 2004.

Bennett, John, and Susan Diana Mary Rowley. *Uqalurait: An Oral History of Nunavut*. Montreal: McGill-Queen's University Press, 2004.

Berkes, Fikret. "Epilogue: Making Sense of Arctic Environmental Change?" In *The Earth Is Faster Now: Indigenous Observations of Arctic Environmental Change*, edited by I. Krupnik and D. Jolly, 334–349. Fairbanks, AK: Arcus, 2002.

———. *Sacred Ecology: Traditional Ecological Knowledge and Resource Management*. Philadelphia, PA: Taylor & Francis, 1999.

Berman, Morris. *Dark Ages America: The Final Phase of Empire*. New York: W. W. Norton and Company, 2006.

Bernstein, Steven. "International Institutions and the Framing of Canada's Climate Change Policy: Mitigating or Masking the Integrity Gap?" In *The Integrity Gap: Canada's Environmental Policy and Institutions*, edited by E. Lee and A. Perl, 68–101. Vancouver: UBC Press, 2003.

Bird-David, Nurit. "Animism Revisited: Personhood, Environment, and Relational Epistemology." *Current Anthropology* 40 (1999): 67–91.

———. "Beyond the Original Affluent Society: A Culturalist Reformulation." *Current Anthropology* 33, no. 1 (1992): 25–47.

Birket-Smith, Kaj. *Ethnography of the Egedesminde District: With Aspects of the General Culture of West Greenland*. New York: AMS Press, 1976.

Black, Richard. "Will Kyoto Die at Canadian Hands?" BBC News, 27 January 2006. http://www.bbc.co.uk.

Blackburn, Carole. *Harvest of Souls: The Jesuit Missions and Colonialism in North America, 1632–1650*. Montreal: McGill-Queen's University Press, 2000.

Bonin, Sister Marie. "The Grey Nuns and the Red River Settlement." *Manitoba History* 11 (Spring 1986). http://www.mhs.mb.ca/docs/mb_history/11/greynuns.shtml.

Borenstein, Seth. "White House Report on Climate Change Has a Warning, and a Silver Lining." *Globe and Mail*, 17 June 2009, sec. A, p. 15.

Boyd, David R., and David Suzuki Foundation. *Sustainability within a Generation: A New Vision for Canada*. Vancouver: David Suzuki Foundation, 2004.

Brady, Maggie. *Heavy Metal: The Social Meaning of Petrol Sniffing in Australia*. Canberra, AU: Aboriginal Studies Press, 1992.

Brody, Hugh. *The Other Side of Eden: Hunters, Farmers and the Shaping of the World*. Vancouver: Douglas & McIntyre, 2000.

———. *Maps and Dreams: Indians and the British Columbia Frontier*. Vancouver: Douglas & McIntyre, 1981.

Brundtland, Gro Harlem, and World Commission on Environment and Development. *Our Common Future*. Oxford: Oxford University Press, 1987.

Buss, Doris, and Didi Herman. *Globalizing Family Values: The Christian Right in International Politics*. Minneapolis: University of Minnesota Press, 2003.

Cairney, Sheree, et al. "The Neurobehavioural Consequences of Petrol (Gasoline) Sniffing." *Neuroscience and Biobehavioral Reviews* 26 (2002): 81–89.

Campbell, Joseph. *Primitive Mythology*. Harmondsworth, UK: Penguin Books, 1976.

Campbell, M., et al. "Dion's Green Plan Would 'Wreak Havoc.'" *Globe and Mail*, 12 September 2008, sec. A, pp. 1, 8.

Canadian Broadcasting Corporation (CBC) News. "Gore: Nobel Prize Win Shows Climate Change a 'Planetary Emergency.'" 12 October 2007. http://www.cbc.ca/world/story/2007/10/12/nobel-peace.html.

———. "Martin Urges Nations to Get Tough on Energy Consumption." 7 December 2005. http://www.cbc.ca/story/canada/national/2005/12/07m artinclimate.

Canadian Marine Environment Protection Society. *Chemistry, Calibre and Climate: The Plight of Canada's Polar Bear*. Vancouver: Canadian Marine Environment Protection Society, 2005.

Carpenter, Joel A. *Revive Us Again: The Reawakening of American Fundamentalism*. New York: Oxford University Press, 1997.

Carpenter, Roger M. *The Renewed, the Destroyed, and the Remade: The Three Thought Worlds of the Huron and the Iroquois, 1609–1650*. East Lansing: Michigan State University Press, 2004.

Chaturvedi, Sanjay. "Arctic Geopolitics Then and Now." In *The Arctic: Environment, People, Policy*, edited by Mark Nuttall and Terry V. Callaghan, 441–457. Amsterdam: Harwood Academic Publishers, 2000.

Christianson, Gale E. *Greenhouse: The 200-Year Story of Global Warming*. New York: Walker and Company, 1999.

Clark, Campbell, and Brian Laghi. "Harper Encouraged by Obama's Climate Policy." *Globe and Mail*, 19 February 2009, sec. A, p. 13.

Cohn, Norman. *Cosmos, Chaos and the World to Come: The Ancient Roots of Apocalyptic Faith*. New Haven: Yale University Press, 1993.

Collier, Michael, and Robert H. Webb. *Floods, Droughts, and Climate Change*. Tucson: University of Arizona Press, 2002.

Commissioner of the Environment and Sustainable Development. *The Commissioner's Perspective 2006: Climate Change, an Overview and Main Points*. Ottawa: Office of the Auditor General of Canada, 2006.

Committee on the Status of Endangered Wildlife in Canada (COSEWIC). *COSEWIC Assessment and Update Status Report on the Polar Bear in Canada*. Ottawa: Environment Canada, 2008.

Cooper, Barry. *It's the Regime, Stupid! A Report from the Cowboy West on Why Stephen Harper Matters*. Toronto: Key Porter Books, 2009.

Crate, Susan A., and Mark Nuttall. "Introduction: Anthropology and Climate Change." In *Anthropology and Climate Change: From Encounters to Actions*, edited by Susan A. Crate and Mark Nuttall, 9–36. Walnut Creek, CA: Left Coast Press, 2009.

Crosby, Alfred W. *Children of the Sun: A History of Humanity's Unappeasable Appetite for Energy*. New York: W.W. Norton, 2006.

Cruikshank, Julie. *Do Glaciers Listen? Local Knowledge, Colonial Encounters, and Social Imagination*. Vancouver: UBC Press, 2005.

———. "Uses and Abuses of 'Traditional Knowledge': Perspectives from the Yukon Territory." In *Cultivating Arctic Landscapes: Knowing and Manag-*

ing Animals in the Circumpolar North, edited by D. G. Anderson and M. Nuttall, 17–32. New York: Berghahn Books, 2004.

Curry, Bill. "Critics Blast Ottawa's 'Shameful' Green Plan." *Globe and Mail*, 20 October 2006, sec. A, pp. 1, 4.

Curry, Bill, and Mark Hume. "PM Plans 'Intensity' Alternative to Kyoto: Blueprint Wouldn't Necessarily Reduce Emissions; Critics React with Scorn." *Globe and Mail*, 11 October 2006, sec. A, pp. 1, 4.

Dateline Hollywood. "Robertson Blames Hurricane on Choice of Ellen Degeneres to Host Emmys." 5 September 2005. http://datelinehollywood. com/archives/2005/09/05/robertson-blames-hurricane-on-choice-of-ellen-degeneres-to-host-emmys/.

David Suzuki Foundation. "New Federal Clean Air Act Won't Clean Air." 19 October 2006. http://www.davidsuzuki.org/climate_change.

———. "News Release: Federal Budget Fails to Mention Global Warming." 2 May 2006. http://www.davidsuzuki.org/climate_change.

———. "Environmental Groups Urge Prime Minister Harper to Meet Kyoto Targets." 12 April 2006. http://www.davidsuzuki.org/climate_change.

Demeritt, David. "The Construction of Global Warming and the Politics of Science." *Annals of the Association of American Geographers* 91, no. 2 (2001): 307–337.

Denov, Myriam, and Kathryn Campbell. "Casualties of Aboriginal Displacement in Canada: Children at Risk Among the Innu of Labrador." *Refuge* 20, no. 2 (2002): 21–33.

Denzin, Norman K., and Yvonna S. Lincoln. "Introduction: The Discipline and Practice of Qualitative Research." In *The Sage Handbook of Qualitative Research*, edited by N. K. Denzin and Y. S. Lincoln, 1–32. Thousand Oaks, CA: Sage Publications, 2005.

Diamond, Jared M. *Collapse: How Societies Choose to Fail or Succeed*. New York: Viking, 2005.

Diamond, Sara. *Spiritual Warfare: The Politics of the Christian Right*. Montreal: Black Rose Books, 1990.

Diet of Souls. Directed by John Houston. Halifax, NS: Triad Films, 2004.

Dion, Stéphane. "Residential Schools Apology." 11 June 2008. http://www. liberal.ca/story_14080_e.aspx.

———. "Canada Will Not Fail the World: An Address by the Honourable Stéphane Dion." Montreal: Liberal Leadership Convention, 1 December 2006.

Doer, Gary, and Jean Charest. "Seize the Climate-Friendly Day." *Globe and Mail*, 7 December 2005, sec. A, p. 27.

Dotto, Lydia. *Storm Warning: Gambling with the Climate of Our Planet*. Toronto: Doubleday Canada, 2000.

Dunlop, S. *A Dictionary of Weather*. Oxford: Oxford University Press, 2001.

Earthtrends. "CO_2: Cumulative Emissions 1990–2000." World Resources Insitutute. http://earthtrends.wri.org (accessed 23 November 2005).

Eccles, W. J. *The French in North America, 1500–1765*. East Lansing: Michigan State University Press, 1998.

Eliade, Mircea. *Shamanism: Archaic Techniques of Ecstasy*. Bollingen Series. Princeton, NJ: Princeton University Press, 1972.

——. *The Myth of the Eternal Return: Or, Cosmos and History*. Bollingen Series. Princeton, NJ: Princeton University Press, 1971.

——. *The Sacred and the Profane: The Nature of Religion*. 1st American ed. New York: Harcourt, Brace, 1959.

Environment Canada. "News Release: Minister Prentice Highlights Progress Made at Polar Bear Roundtable." 16 January 2009. http://www.ec.gc.ca/default.asp?lang=En&n=714D9AAE-1&news=24AABBD9-00C3-4E80-9517-2D37013C5FAF.

Ernsting, Michele. "The Meaning of Sila." *Radio Netherlands*, 21 December 2001.

Essex, Christopher, and Ross McKitrick. *Taken by Storm: The Troubled Science, Policy and Politics of Global Warming*. Toronto: Key Porter Books, 2002.

Esteva, Gustavo. "Development." In *The Development Dictionary: A Guide to Knowledge as Power*, edited by W. Sachs, 6–25. London: Zed Books, 1992.

Favret-Saada, Jeanne. *Deadly Words: Witchcraft in the Bocage*. Cambridge, UK: Cambridge University Press, 1980.

Fekete, Jason, and Don Braid. "Stelmach Tories Praise Ignatieff over Harper." *Calgary Herald*, 9 June 2009.

Ferland-Angers, Albertine, and Grey Nuns. *Mother d'Youville: First Canadian Foundress: Marie-Marguerite Du Frost De Lajemmerais, Widow d'Youville, 1701–1771: Foundress of the Sisters of Charity of the General Hospital of Montreal, Grey Nuns*. Montreal: Sisters of Charity of Montreal "Grey Nuns," 2000.

Fitts, Mary Pauline. *Hands to the Needy: Mother d'Youville, Apostle to the Poor*. 1st ed. Garden City, NY: Doubleday, 1950.

Flannery, Tim F. *The Weather Makers: How We Are Changing the Climate and What It Means for Life on Earth*. 1st Canadian ed. Toronto: HarperCollinsCanada, 2006.

Fleming, James Rodger. *Historical Perspectives on Climate Change*. New York: Oxford University Press, 1998.

Fox, Shari. "These Are Things That Are Really Happening: Inuit Perspectives on the Evidence and Impacts of Climate Change in Nunavut." In *The Earth Is Faster Now: Indigenous Observations of Arctic Environmental Change*, edited by I. Krupnik and D. Jolly, 12–53. Fairbanks, AK: Arcus, 2002.

Friesen, John W. *Aboriginal Spirituality and Biblical Theology: Closer than You Think*. Calgary: Detselig, 2000.

Galloway, Gloria. "PM Gives Strong Defence of Energy Sector." *Globe and Mail*, 20 October 2006, sec. A, p. 4.

Gardner, Gary. "Engaging Religion in the Quest for a Sustainable World." In *State of the World 2003*, edited by L. Starke, 152–175. New York: W. W. Norton and Company.

George, Jane. "ICC Moves to Patch Up Inuit Climate Change Rift: There Is a

Paradox of Development Among Inuit." *Nunatsiaq News*, 17 December 2009. http://www.nunatsiaqonline.ca.

———. "Inuit Leaders at Odds over Oil and Gas Emissions: Greenland Wants More Development of Carbon-spewing Industries." *Nunatsiaq News*, 13 December 2009. http://www.nunatsiaqonline.ca.

Gibbins, Roger, and Sonia Arrison. *Western Visions: Perspectives on the West in Canada*. Peterborough, ON: Broadview Press, 1995.

Gitay, Habiba, et al. "Ecosystems and Their Goods and Services." In *Climate Change 2001: Impacts, Adaptation and Vulnerability*, contribution of Working Group II to the TAR of the IPCC, 237–342. Cambridge: Cambridge University Press, 2001.

Good News Bible: The Bible in Today's English Version. Toronto: Canadian Bible Society, 1976.

Government of Canada. *Moving Forward on Climate Change: A Plan for Honouring Our Kyoto Commitment*. 2005. http://www.climatechange.gc.ca.

Gray, John. "America's Global Fall from Grace." *Globe and Mail*, 1 October 2008, sec. A, p. 21.

Hallman, David G. "Climate Change: Ethics, Justice, and Sustainable Community." In *Christianity and Ecology: Seeking Well-being of Earth and Humans*, edited by D. T. Hessel and R. R. Ruether, 453–471. Cambridge: Harvard University Press, 2000.

———. "Ethics and Sustainable Development." In *Ecotheology: Voices from South and North*, edited by D. G. Hallman, 264–284. Maryknoll, NY: Orbis Books, 1994.

Harper, Prime Minister Stephen. "Dion's Carbon Tax Flip-Flop Will Punish Taxpayers." 19 June 2008. http://www.conservative.ca/EN/1091/100824.

———. "Prime Minister Harper Offers Full Apology on Behalf of Canadians for the Indian Residential Schools System." 11 June 2008. http://www.conservative.ca/EN/1091/100669.

———. "Prime Minister Harper Bolsters Arctic Sovereignty with Science and Infrastructure Announcements." Churchill: MB, 5 October 2007. http://pm.gc.ca/eng/media.asp?id=1843.

———. "Harper's Index: Stephen Harper Introduces the Tar Sands Issue." 14 July 2006. http://www.dominionpaper.ca/articles/1491.

Hassol, Susan Joy, et al. *Impacts of a Warming Arctic: Arctic Climate Impact Assessment*. Cambridge: Cambridge University Press, 2004.

Hawken, Paul, L. Hunter Lovins, and Amory B. Lovins. *Natural Capitalism: Creating the Next Industrial Revolution*. Boston: Little, Brown and Co., 1999.

Hay, Colin. "Environmental Security and State Legitimacy." In *Is Capitalism Sustainable? Political Economy and the Politics of Ecology*, edited by M. O'Connor, 217–231. New York: The Guilford Press, 1994.

Hessing, M., M. Howlett, and T. Summerville. *Canadian Natural Resource and Environmental Policy*. 2nd ed. Vancouver: UBC Press, 2005.

Homer-Dixon, Thomas. *The Ingenuity Gap*. Toronto: Vintage Canada, 2001.

Homer-Dixon, Thomas, and Andrew Weaver. "Responding to the Skeptics." *Globe and Mail*, 7 December 2009, sec. A, p. 15.

Howlett, Karen, S. Chase, and D. Leblanc. "Battle Begins for Elusive Majority." *Globe and Mail*, 8 September 2008, sec. A, pp. 1, 4.

Hulan, Renée. *Northern Experience and the Myths of Canadian Culture.* Montreal: McGill-Queen's University Press, 2002.

Hultkrantz, Åke. "Les Religions du Grand Nord Américain." In *Les Religions Arctiques et Finnoises*, edited by I. Paulson, Å Hultkrantz, and K. Jettmar, 341–395. Paris: Payot, 1965.

Hussen, Ahmed M. *Principles of Environmental Economics: Economics, Ecology and Public Policy.* London: Routledge, 1999.

Ignatieff, Michael. *True Patriot Love: Four Generations in Search of Canada.* Toronto: Penguin Canada, 2009.

———. "Statement from Liberal Leader Michael Ignatieff on International Earth Day." 22 April 2009. http://www.liberal.ca/en/newsroom/media-releases/15747_statement-from-liberal-leader-michael-ignatieff-on-international-earth-day.

———. "Statement from the Liberal Leader Michael Ignatieff on the First Anniversary of the Residential Schools Apology." 11 June 2009. http://www.liberal.ca/en/newsroom/media-releases/15921_statement-from-liberal-leader-michael-ignatieff-on-the-first-anniversary-of-the-residential-schools-apology.

———. *The Rights Revolution.* Toronto: Anansi, 2000.

Immerwahr, John. *Waiting for a Signal: Public Attitudes Toward Global Warming, the Environment, and Geophysical Research.* New York: Public Agenda, 1999. http://Earth.agu.org/sci_soc.html, 1999.

Ingold, Tim. "Rethinking the Animate, Re-Animating Thought." *Ethnos* 71, no. 1 (2006): 9–20.

———. "On the Social Relations of the Hunter-Gatherer Band." In *The Cambridge Encyclopedia of Hunters and Gatherers*, edited by R. B. Lee and R. Daly, 399–410. Cambridge: Cambridge University Press, 1999.

Intergovernmental Panel on Climate Change (IPCC). *Climate Change 2007: The Physical Science Basis, Summary for Policymakers, Contribution of Working Group I to the Fourth Assessment Report of the IPCC.* Geneva, Switzerland: IPCC Secretariat/World Meteorological Organization, 2007.

———. *Climate Change 2007:Climate Change Impacts, Adaptation, and Vulnerability, Summary for Policymakers, Contribution of Working Group II to the Fourth Assessment Report of the IPCC.* Geneva, Switzerland: IPCC Secretariat/World Meteorological Organization, 2007.

———. *Climate Change 2001: Impacts, Adaptations, and Vulnerability: Summary for Policymakers, a Report of Working Group II of the Intergovernmental Panel on Climate Change and Technical Summary of the Working Group II Report, a Report Accepted by Working Group II of the IPCC.* Geneva, Switzerland: IPCC Secretariat/World Meteorological Organization, 2001.

———. *Climate Change 2001: Synthesis Report: Contributions of Working Group I, II, and III to the Third Assessment Report of the IPCC*. Geneva, Switzerland: IPCC Secretariat/World Meteorological Organization, 2001.

———. *IPCC Special Report: Emissions Scenarios, a Report of IPCC Working Group III*. Geneva, Switzerland: IPCC Secretariat/World Meteorological Organization, 2000.

Inuit Circumpolar Conference. "Inuit Petition Inter-American Commission on Human Rights to Oppose Climate Change Caused by the United States of America. 7 December 2005. http://www.inuitcircumpolar.com.

Inuit Tapiriit Kanatami and Inuit Circumpolar Council. "Inuit of Canada Expect Substantial Consultations Prior to Canadian Polar Bear Listing, Joint Media Release." 20 January 2009. http://www.itk.ca/media-centre/media-releases/inuit-canada-expect-substantial-consultations-prior-canadian-polar-bear-.

Inuktitut Living Dictionary. 12 July 2006. http://livingdictionary.com.

Jaccard, Mark, John Nyboer, and Bryn Sadownik. *The Cost of Climate Policy*. Vancouver: UBC Press, 2002.

Jolly, Dyanna, et al. "We Can't Predict the Weather Like We Used To." In *The Earth Is Faster Now: Indigenous Observations of Arctic Environmental Change*, edited by I. Krupnik and D. Jolly, 93–125. Fairbanks, AK: Arcus, 2002.

Keller, Catherine. *Apocalypse Now and Then: A Feminist Guide to the End of the World*. Boston: Beacon Press, 1996.

Kennedy, Robert Francis. *Crimes Against Nature: How George W. Bush and His Corporate Pals Are Plundering the Country and High-Jacking Our Democracy*. 1st ed. New York: HarperCollins, 2004.

Kidwell, Clara Sue, Homer Noley, and George E. Tinker. *A Native American Theology*. Maryknoll, NY: Orbis Books, 2001.

Kirchner, J. W. "The Gaia Hypothesis: Conjectures and Refutations." *Climatic Change* 58 (2003): 21–45.

———. "The Gaia Hypothesis: Facts, Theory, and Wishful Thinking." *Climatic Change* 52 (2002): 391–408.

Kleidon, A. "Testing the Effect of Life on Earth's Functioning: How Gaia Is the Earth System?" *Climatic Change* 52 (2002): 383–389.

Laidler, Gita J. "Inuit and Scientific Perspectives on the Relationship between Sea Ice and Climate Change: The Ideal Complement." *Climatic Change* 78 (2006): 407–444.

Lasch, Christopher. *The True and Only Heaven: Progress and Its Critics*. New York: Norton, 1991.

Latour, Bruno. *We Have Never Been Modern*. Translated by C. Porter. Cambridge: Harvard University Press, 1993.

Lee, Eugene, and Anthony Perl. "Introduction: Institutions and the Integrity Gap in Canadian Environmental Policy." In *The Integrity Gap: Canada's Environmental Policy and Institutions*, edited by E. Lee and A. Perl, 3–24. Vancouver: UBC Press, 2003.

Lemmen, Donald Stanley, Fiona J. Warren, and Climate Change Impacts and Adaptation Program. *Climate Change Impacts and Adaptation: A Canadian Perspective*. Ottawa: Climate Change Impacts and Adaptation Program, 2004.

Lenton, Timothy M. "Testing Gaia: The Effect of Life on Earth's Habitability and Regulation." *Climatic Change* 52 (2002): 409–422.

Lenton, Timothy M., and David M. Wilkinson. "Developing the Gaia Theory: A Response to the Criticisms of Kirchner and Volk." *Climatic Change* 58 (2003): 1–12.

Liberal Party. *The Green Shift*. http://www.thegreenshift.ca/default_e.aspx (accessed 17 August 2008).

———. *The Green Shift: Building a Canadian Economy for the Twenty-First Century*. 2008. http://www.cbc.ca/newsatsixns/pdf/liberalgreenplan.pdf.

Livingston, John A. *The John A. Livingston Reader: The Fallacy of Wildlife Conservation and One Cosmic Instant*. Toronto: McClelland and Stewart, 2007.

———. *Rogue Primate: An Exploration of Human Domestication*. Toronto: Key Porter Books, 1994.

———. *Arctic Oil*. Toronto: CBC Merchandising, 1981.

Lopez, Barry. *Arctic Dreams: Imagination and Desire in a Northern Landscape*. Toronto: Bantam Books, 1989.

Lovelock, James E. "Gaia and Emergence: A Response to Kirchner and Volk." *Climatic Change* 57 (2003): 1–3.

———. *Gaia, a New Look at Life on Earth*. Oxford: Oxford University Press, 1979.

Macdougall, Doug. *Frozen Earth: The Once and Future Story of Ice Ages*. Berkeley: University of California Press, 2004.

Malin, Shimon. *Nature Loves to Hide: Quantum Physics and Reality, a Western Perspective*. Oxford: Oxford University Press, 2001.

Mandeville, Bernard. *The Fable of the Bees, Or, Private Vices, Publick Benefits*. Indianapolis: Liberty Classics, 1988.

Martin, Paul. "Address by Prime Minister Paul Martin at the UN Conference on Climate Change." Montreal, QC, 7 December 2005. http://pm.gc.ca/eng/news.asp?id=666.

Mason, Gary. "The Environment Was not a Winning Issue on this Campaign Trail." *Globe and Mail*, 15 October 2008, sec. A, p. 7.

Matthiasson, John Stephen. *Living on the Land: Change among the Inuit of Baffin Island*. Peterborough, ON: Broadview Press, 1992.

McCarthy, Shawn. "Canada's Strategy: Promise Now, Implement Later." *Globe and Mail*, 19 December 2009, sec. A, p. 18.

———. "A Red Flag in the Global-Warming Battle." *Globe and Mail*, 11 October 2006, sec. A, p. 4.

McCarthy, Shawn, and C. Clark. "Ottawa Swoops in with Climate-Change Offer." *Globe and Mail*, 6 November 2008, sec. A, pp. 1, 8.

McGhee, Robert. *Ancient People of the Arctic*. Vancouver: UBC Press, 1996.

McGrath, Alister E. *The Reenchantment of Nature: The Denial of Religion, and the Ecological Crisis.* 1st ed. New York: Doubleday, 2002.

McKenna, Barrie. "Greenspan Admits 'Mistake' on Bank Regulation." *Globe and Mail,* 24 October 2008, sec. A, pp. 1, 16.

McKenzie, Judith. *Environmental Politics in Canada: Managing the Commons into the Twenty-First Century.* Don Mills, ON: Oxford University Press, 2002.

McKibben, Bill. "The Christian Paradox: How a Faithful Nation Gets Jesus Wrong." *Harper's Magazine,* August 2005, 31–37.

McKitrick, Ross. *An Economist's Perspective on Climate Change and the Kyoto Protocol.* Presentation to the Department of Economics Annual Fall Workshop. The University of Manitoba, Winnipeg, MB. 7 November 2003.

McNeill, John Robert. *Something New under the Sun: An Environmental History of the Twentieth-Century World.* 1st ed. New York: W. W. Norton and Company, 2000.

McQuaig, Linda. *It's the Crude, Dude: War, Big Oil and the Fight for the Planet.* Toronto: Doubleday Canada, 2004.

Mendel, Arthur P. *Vision and Violence.* Ann Arbor: University of Michigan Press, 1999.

Merkur, Daniel. *Powers Which We Do Not Know: The Gods and Spirits of the Inuit.* Moscow, ID: University of Idaho Press, 1991.

———. "Breath-soul and the Wind Owner: The Many and the One in Inuit Religion." *American Indian Quarterly* 7, no. 3 (1983): 23–39.

Miller, J. R. *Skyscrapers Hide the Heavens: A History of Indian-White Relations in Canada.* 3rd ed. Toronto: University of Toronto Press, 2000.

Minowitz, Peter. *Profits, Priests, and Princes: Adam Smith's Emancipation of Economics from Politics and Religion.* Stanford, CA: Stanford University Press, 1993.

Mittelstaedt, Martin. "Aviation Industry in Eye of Climate-Change Storm." *Globe and Mail,* 14 May 2007, sec. A, p. 6.

———. "What Does Ottawa's Green Plan Entail?" *Globe and Mail,* 20 October 2006, sec. A, p. 5.

———. "Ottawa Stops Funding One Tonne Challenge." *Globe and Mail,* 1 April 2006, sec. A, p. 7.

Monbiot, George. "Canada's Image Lies in Tattters: It Is Now to Climate What Japan Is to Whaling." *The Guardian,* 30 November 2009. http://guardian.co.uk.

Montgomery, Charles. "Nurturing Doubt about Climate Change Is Big Business." *Globe and Mail,* 12 August 2006, sec. F, pp. 4–5.

Moogk, Peter N. *La Nouvelle France: The Making of French Canada: A Cultural History.* East Lansing: Michigan State University Press, 2000.

Moyers, Bill. "Welcome to Doomsday." *The New York Review of Books,* 24 March 2005, 52, 5.

Munro, Margaret. "Polar Bears Face Uncertain Future in Arctic." *Dose,* 28 December 2008. http://www.dose.ca.

Murphy, Rex. "Through Copenhagen's Looking Glass." *Globe and Mail*, 19 December 2009, sec. A, p. 25.

———. "Be Brave: Escape the Climate Box." *Globe and Mail*, 17 October 2009, sec. A, p. 27.

Murphy, Sister Mary. "The Grey Nuns Travel West." *Manitoba Historical Society Transaction Series*, December 1944, 3. http://www.mhs.mb.ca/docs/transactions/3/ greynuns.shtml.

National Association of Evangelicals. *For the Health of the Nation: An Evangelical Call to Civic Responsibility*. http://www.nae.net (accessed 16 July 2006).

National Post. "Open Kyoto to Debate: Sixty Scientists Call on Harper to Revisit the Science of Global Warming." 6 April 2006.

Nelson, Richard K. *Make Prayers to the Raven: A Koyukon View of the Northern Forest*. Chicago: The University of Chicago Press, 1983.

Nelson, Robert H. *Economics as Religion: From Samuelson to Chicago and Beyond*. University Park, PA: Pennsylvania State University Press, 2001.

———. *Reaching for Heaven on Earth: The Theological Meaning of Economics*. Savage, MD: Rowman & Littlefield Publishers, 1991.

Nikiforuk, Andrew. *Tar Sands: Dirty Oil and the Future of a Continent*. Vancouver: GreyStone Books, 2008.

Northcott, Michael S. *A Moral Climate: The Ethics of Global Warming*. London: Darton, Longman and Todd, 2007.

Nuffield, E. W. *The Discovery of Canada*. Vancouver: Haro Books, 1996.

Nuliajuk: Mother of the Sea Beasts. Directed by John Houston. Halifax, NS: Triad Films, 2001.

Nuttall, Mark. "Indigenous Peoples, Self-determination and the Arctic Environment." In *The Arctic: Environment, People, Policy*, edited by M. Nuttall and T. V. Callaghan, 377–409. Amsterdam: Harwood Academic Publishers, 2000.

———. "Indigenous Peoples' Organizations and Arctic Environmental Cooperation." In *The Arctic: Environment, People, Policy*, edited by M. Nuttall and T. V. Callaghan, 621–636. Amsterdam: Harwood Academic Publishers, 2000.

Nuttall, Mark, and Terry V. Callaghan. "Introduction." In *The Arctic: Environment, People, Policy*, edited by Mark Nuttall, Terry V. Callaghan, xxv-xxxviii. Amsterdam: Harwood Academic Publishers, 2000.

Ó Crualaoich, Gearóid. *The Book of The Cailleach: Stories of the Wise-Woman Healer*. Cork, Ireland: Cork University Press, 2003.

O'Leary, Stephen D. *Arguing the Apocalypse: A Theory of Millennial Rhetoric*. New York: Oxford University Press, 1994.

Olsen, Jr., Richard K. "Living Above It All: The Liminal Fantasy of Sport Utility Vehicle Advertisements." In *Enviropop: Studies in Environmental Rhetoric and Popular Culture*, edited by M. Meister and P. M. Japp, 175–196. Westport, CT: Praeger, 2002.

Pelly, David F. *Sacred Hunt: A Portrait of the Relationship between Seals and Inuit*. Vancouver: Douglas & McIntyre, 2001.

Perl, Anthony, and Eugene Lee. "Conclusion." In *The Integrity Gap: Canada's Environmental Policy and Institutions*, edited by E. Lee and A. Perl, 241–270. Vancouver: UBC Press, 2003.

Perlez, Jane. "U.S. to Join China and India in Climate Pact." *New York Times*, 27 July 2005. http://www.nytimes.com.

Petrone, Penny. *Northern Voices: Inuit Writing in English*. Toronto: University of Toronto Press, 1988.

Pettazzoni, Raffaele. *The All-knowing God*. London: Methuen and Co., Ltd., 1956.

Philander, S. George. *Is the Temperature Rising? The Uncertain Science of Global Warming*. Princeton, NJ: Princeton University Press, 1998.

Phillips, Kevin P. *American Theocracy: The Peril and Politics of Radical Religion, Oil, and Borrowed Money in the 21st Century*. New York: Viking, 2006.

Plumwood, Val. *Environmental Culture: The Ecological Crisis of Reason*. London: Routledge, 2002.

———. "Human Vulnerability and the Experience of Being Prey." *Quadrant* 39, no. 314 (1995): 29–34.

Polanyi, Karl. *The Great Transformation*. Boston: Beacon Press, 1957.

Polar Bears International. "National Roundtable on Polar Bears, Media Release." 16 January 2009. http://www.polarbearsinternational.org/in-the-news/polar-bear-roundtable/.

Primavesi, Anne. *Gaia and Climate Change: A Theology of Gift Events*. London: Routledge/Taylor & Francis Group, 2009.

———. *Gaia's Gift: Earth, Ourselves and God After Copernicus*. London: Routledge/Taylor & Francis Group, 2003.

———. *Sacred Gaia: Holistic Theology and Earth System Science*. London: Routledge, 2000.

Qitsualik, Rachel Attituq. "Sila." *Nunatsiaq News*, 7 July 2000. http://www.nunatsiaqonline.ca/.

———. "Word and Will, Part Two: Words and the Substance of Life." *Nunavut Edition*, 12 November 1998.

Rappaport, Roy A. *Ritual and Religion in the Making of Humanity*. Cambridge: Cambridge University Press, 1999.

Rasmussen, Knud. *Across Arctic America: Narrative of the Fifth Thule Expedition*. New York: Greenwood Press, 1969.

Rasmussen, Knud, and Hother Berthel Simon Ostermann. *The Alaskan Eskimos as Described in the Posthumous Notes of Knud Rasmussen*. Report of the Fifth Thule Expedition, 1921–1924. Vol. 10, no. 3. Copenhagen: Gyldendal, 1952.

Reguly, Eric, and Shawn McCarthy. "The Copenhagen Summit: Agreement on Climate Reached—but Unfinished." *Globe and Mail*, 19 December 2009, sec. A, pp. 1, 22.

Remie, Cornelius H. W., and Jarich Oosten. "The Birth of a Catholic Inuit

Community: The Transition to Christianity in Pelly Bay, Nunavut, 1935–1950." *Études/Inuit/Studies* 26, no. 1 (2002): 109–141.

Repent America. "Hurricane Katrina Destroys New Orleans Days before 'Southern Decadence.'" 31 August 2005. http://www.repentamerica.com/pr_hurricanekatrina.html.

Richling, Barnett. "Very Serious Reflections: Inuit Dreams About Salvation and Loss in Eighteenth-Century Labrador." *Ethnohistory* 36, no. 2 (1989): 148–169.

Riedlinger, D., and F. Berkes. "Contributions of Traditional Knowledge to Understanding Climate Change in the Canadian Arctic." *Polar Record* 37, no. 203 (2001): 315–328.

Rist, Gilbert. *The History of Development: From Western Origins to Global Faith.* London: Zed Books, 2002.

Robbins, Lionel. *An Essay on the Nature & Significance of Economic Science.* London: Macmillan, 1952.

Roberts, Paul. *The End of Oil: On the Edge of a Perilous New World.* Boston: Houghton, 2004.

Robertson, Pat. "Letter: Pat Robertson Corrects Dateline Hollywood Article." 18 September 2005. http://datelinehollywood.com/archives/2005/09/18/pat-robertson-corrects-dateline-hollywood-article/.

———. *The New World Order.* Dallas: Word Pub., 1991.

Rockefeller, Steven C. "Global Interdependence, the Earth Charter, and Christian Faith." In *Earth Habitat: Eco-Injustice and the Church's Response,* edited by D. Hessel and L. Rasmussen, 101–121. Minneapolis: Fortress Press, 2001.

Rogers, Raymond A. "The Usury Debate, the Sustainability Debate, and the Call for a Moral Economy." *Ecological Economics* 35 (2000): 157–171.

———. *Solving History: The Challenge of Environmental Activism.* Montreal: Black Rose Books, 1998.

———. *Nature and the Crisis of Modernity: A Critique of Contemporary Discourse on Managing the Earth.* Montreal: Black Rose Books, 1994.

Rogers, Raymond A., et al. "The Why of the 'Hau': Scarcity, Gifts, and Environmentalism." *Ecological Economics* 51 (2004): 177–189.

Root, Terry L., and Stephen H. Schneider. "Climate Change: Overview and Implications for Wildlife." In *Wildlife Responses to Climate Change: North American Case Studies,* edited by S. H. Schneider and T. L. Root, 1–56. Washington, DC: Island Press, 2002.

Rothschild, Emma. *Economics Sentiments: Adam Smith, Condorcet, and the Enlightenment.* Cambridge: Harvard University Press, 2001.

Royal Commission on Aboriginal Peoples. *Report of the Royal Commission on Aboriginal Peoples.* Ottawa: Indian and Northern Affairs Canada. http://www.ainc-inac.gc.ca/ch/rcap/sg/sgmm_e.html (accessed 10 November 2005).

Ruddiman, William F. *Plows, Plagues, and Petroleum: How Humans Took Control of Climate.* Princeton, NJ: Princeton University Press, 2005.

Sachs, Wolfgang. "Environment." In *The Development Dictionary*, edited by W. Sachs, 26–37. London: Zed Books, 1992.

Sahlins, Marshall. "The Sadness of Sweetness: The Native Anthropology of Western Cosmology." *Current Anthropology* 37, no. 3 (1996): 395–428.

———. *Stone Age Economics*. London: Tavistock, 1972.

Sankar, U. *Environmental Economics*. Oxford: Oxford University Press, 2002.

Saunders, Doug. "Copenhagen Summit: Breach in the Global-Warming Bunker Rattles Climate Science at Worse Time." *Globe and Mail*, 5 December 2009, sec. A, pp. 1, 22.

Sbert, José M. "Progress." In *The Development Dictionary: A Guide to Knowledge as Power*, edited by W. Sachs, 192–205. London: Zed Books, 1992.

Scharper, Stephen. "The Gaia Hypothesis: Implications for a Christian Political Theology of the Environment." *Cross Currents* 44, no. 2 (1994): 207–221.

Scherer, Glenn. "The Godly Must Be Crazy: Christian-Right Views Are Swaying Politicians and Threatening the Environment." *Grist Magazine: Environmental News and Commentary*, 27 October 2004. http://www.grist.org/news/maindish/2004/10/27/scherer-christian.

Schmidt, Wilhelm. *The Origin and Growth of Religion: Facts and Theories*. New York: Cooper Square Publishers, Inc., 1972.

Schneider, Stephen, and Jose Sarukhan. "Overview of Impacts, Adaptation, and Vulnerability to Climate Change." In *Climate Change 2001: Impacts, Adaptation, and Vulnerability*, contribution of Working Group II to the TAR of the IPCC, 77–103. Cambridge: Cambridge University Press, 2001.

Schröder, Heike. *Negotiating the Kyoto Protocol: An Analysis of Negotiation Dynamics in International Negotiations*. Münchener Beiträge Zur Geschichte Und Gegenwart Der Internationalen Politik. Münster: Transaction Publishers, 2001.

Schwartz, Peter, and Doug Randall. *An Abrupt Climate Change Scenario and Its Implications for United States National Security*. Washington, DC: Pentagon Briefing Paper, October 2003.

Scott, G. Richard, et al. "Physical Anthropology of the Arctic." In *The Arctic: Environment, People, Policy*, edited by M. Nuttall and T. V. Callaghan, 339–373. Amsterdam: Harwood Academic Publishers, 2000.

Segal, Bernard. "The Inhalant Dilemma: A Theoretical Perspective." In *Sociocultural Perspectives on Volatile Solvent Use*, edited by F. Beauvais and J. E. Trimble, 79–102. New York: The Haworth Press, Inc., 1997.

Siegert, Martin J., and Julian A. Dowdeswell. "Glaciology." In *The Arctic: Environment, People, Policy*, edited by M. Nuttall and T. V. Callaghan, 27–55. Amsterdam: Harwood Academic Publishers, 2000.

Sierra Club of Canada. *No More Idling: California Standards Needed Now!* 19 October 2006. http://www.sierraclub.ca.

———. "Canadian Climate Coalition Denounces Conservative Party for Ducking the Issues." 17 January 2006. http://www.sierraclub.ca.

————. "Harper's Position on Kyoto: A Tragedy for the Planet." 12 January 2006. http://www.sierraclub.ca.

Sila Alangotok. Directed by Bonnie Dickie, et al. Winnipeg, MB: International Institute for Sustainable Development, 2000.

Simpson, Jeffrey. "Little New for Obama in Ottawa's Energy 'Offer.'" *Globe and Mail,* 12 November 2008, sec. A, p. 23.

————. "When All's Said and Done the Carbon Tax Is Toast." *Globe and Mail,* 22 October 2008, sec. A, p. 21.

Singer, Peter. *The President of Good & Evil: The Ethics of George W. Bush.* New York: Dutton, 2004.

Skogan, Joan. *Mary of Canada: The Virgin Mary in Canadian Culture, Spirituality, History, and Geography.* Banff, AB: Banff Centre Press, 2003.

Smit, Barry, and Olga Pilifosova. "Adaptation to Climate Change in the Context of Sustainable Development and Equity." In *Climate Change 2001: Impacts, Adaptation, and Vulnerability,* contribution of Working Group II to the TAR of the IPCC, 879–912. Cambridge: Cambridge University Press, 2001.

Smith, Linda Tuhiwai. *Decolonizing Methodologies: Research and Indigenous Peoples.* London: Zed Books, 1999.

Sopinka, Heidi. "Carbon Offsets: A Shell Game?" *Globe and Mail,* 30 May 2007, sec. L, pp. 1–2.

Statement of the Evangelical Climate Initiative. *Climate Change: An Evangelical Call to Action.* 2006. http://www.christiansandclimate.org.

Stephen, Michele. *A'aisa's Gifts: A Study of Magic and the Self.* Berkeley, CA: University of California Press, 1995.

Suzuki, David, with Faisal Moola. "Canada Must Do More to Confront Climate Crisis." *Canoe Network,* 20 November 2009. http://cnews.canoe.ca.

Taber, Jane. "Abstaining No Longer a Liberal Option." *Globe and Mail,* 24 October 2008, sec. A, p. 4.

————. "Dion Ignored Advisers' Advice, Preferring to Act as 'a Lone Wolf.'" *Globe and Mail,* 17 October 2008, sec. A, p. 9.

————. "Liberals Hit Turbulence as Campaign Takes Off." *Globe and Mail,* 8 September 2008, sec. A, pp. 1, 8.

Tainter, Joseph A. "Global Change, History, and Sustainability." In *The Way the Wind Blows: Climate, History, and Human Action,* edited by R. J. McIntosh, J. A. Tainter, and S. K. McIntosh, 331–353. New York: Columbia University Press, 2000.

————. *The Collapse of Complex Societies.* New Studies in Archaeology. Cambridge: Cambridge University Press, 1988.

Taylor, Sarah McFarland. *Green Sisters: A Spiritual Ecology.* Cambridge: Harvard University Press, 2007.

Tertzakian, Peter. *A Thousand Barrels a Second: The Coming Oil Break Point and the Challenges Facing an Energy Dependent World.* New York: McGraw-Hill, 2006.

Thorpe, Natasha, et al. "Nowadays It Is Not the Same." In *The Earth Is Faster Now: Indigenous Observations of Arctic Environmental Change*, edited by I. Krupnik and D. Jolly, 201–239. Fairbanks, AK: Arcus, 2002.

Tinker, George E. "Community and Ecological Justice: A Native American Response." In *Earth at Risk: An Environmental Dialogue Between Religion and Science*, edited by D. B. Conroy and R. L. Petersen, 239–258. Amherst, NY: Humanity Books, 2000.

Tol, Richard S. "Is the Uncertainty about Climate Change Too Large for Expected Cost-Benefit Analysis?" *Climatic Change* 56 (2003): 265–289.

Torrance, Robert M. *The Spiritual Quest: Transcendence in Myth, Religion, and Science*. Berkeley: University of California Press, 1994.

Tylor, Edward Burnett. *Primitive Culture*. New York: Harper, 1958.

United Nations Framework Convention on Climate Change. *Key GHG Data*. 2005. http://unfccc.int.2860.php.

Volk, Tyler. "Toward a Future for Gaia Theory: An Editorial Comment." *Climatic Change* 52 (2002): 423–430.

Watt-Cloutier, Sheila, Terry Fenge, and Paul Crowley. *Responding to Global Climate Change: The Perspective of the Inuit Circumpolar Conference on the Arctic Climate Impact Assessment*. 16 February 2005. http://www.inuitcircumpolar.com.

Weart, Spencer R. *The Discovery of Global Warming*. New Histories of Science, Technology, and Medicine. Cambridge: Harvard University Press, 2003.

Wente, Margaret. "Why People Are Chilled by Warming." *Globe and Mail*, 15 October 2009, sec. A, p. 21.

Wenzel, George. "From TEK to IQ: Inuit Qaujimajatuqangit and Inuit Cultural Ecology." *Arctic Anthropology* 41, no. 2 (2004): 238–250.

———. "Polar Bear as Resource." Yellowknife: Paper delivered at the 3rd Northern Research Forum (NRF) Open Meeting. Yellowknife, NT. 2004.

———. "Ninqiqtuq: Inuit Resource Sharing and Generalized Reciprocity in Clyde River, Nunavut." *Arctic Anthropology* 32, no. 2 (1995): 43–60.

———. *Animal Rights, Human Rights: Ecology, Economy and Ideology in the Canadian Arctic*. London: Belhaven Press, 1991.

White, Jr., Lynn. "The Historical Roots of Our Ecologic Crisis." *Science* 155, no. 3767 (1967): 1203–1207.

Wilson, Bruce. "Pat Robertson's Sweaty Global Warming Epiphany Challenges American Environmental Movement." 5 August 2006. http://www.talk2action.org.

Wohlforth, Charles P. *The Whale and the Supercomputer: On the Northern Front of Climate Change*. 1st ed. New York: North Point Press, 2004.

Wójcik, Daniel. *The End of the World As We Know It: Faith, Fatalism, and Apocalypse in America*. New York: New York University Press, 1997.

Wood, Ellen Meiksins. *The Origin of Capitalism*. New York: Monthly Review Press, 1999.

World Net Daily. "A Green Gospel: Pat Robertson Converts—to 'Global Warming.'" 3 August 2006. http://www.WorldNetDaily.com.

Xenos, Nicholas. *Scarcity and Modernity*. London: Routledge, 1989.

York, Geoffrey. "Draft Bali Deal Omits Specific Targets." *Globe and Mail*, 15 December 2007, sec. A, p. 1.

———. "Baird a No-Show at Key Negotiating Session." *Globe and Mail*, 15 December 2007, sec. A, p. 2.

———. "Business Gets a Voice on Canadian Delegation." *Globe and Mail*, 10 December 2007, sec. A, p. 17.

Zavaleta, Erika S., and Jennifer L. Royval. "Climate Change and the Susceptibility of U.S. Ecosystems to Biological Invasions: Two Cases of Expected Range Expansion." In *Wildlife Responses to Climate Change: North American Case Studies*, edited by S. H. Schneider and T. L. Root, 277–342. Washington, DC: Island Press, 2002.

Riedlinger, Dyanna, 4, 20, 23, 100
Rio de Janeiro Earth Summit, 82,
   83, 226
Robbins, Lionel, 124
Roberts, Paul, 161–162, 166
Robertson, Pat, 149, 150, 152, 167,
   168–169
Rogers, Raymond, 125, 127–128
Royal Commission on Aboriginal
   Peoples, 25, 116, 117
Ruddiman, William F., 55, 56

S
Sachs, Wolfgang, 50, 59, 60, 67, 82
Sahlins, Marshall, 121–122, 123, 191
Sammurtok, Simionie, 20, 27, 185–
   186, 213, 231
scarcity, law of, 124, 129
Scharper, Stephen, 96
Scherer, Glenn, 166
Schwartz, Peter, 160–161, 164
scientific principles, uncertainty, 62,
   65–66, 87, 218–219
sea ice changes, 24
seal hunt debate, 3–4
Sedna
animism, 183, 185–186
hunting and, 189–191, 192, 193, 194,
   195–196
shamans and, 203–204
significance of, 182–183, 187, 195, 203
stories of, 180, 181–182, 195–196
taboo systems, 194–197
self-representation, Inuit, 35–36
sexual abuses in residential schools,
   116–117
shamanism
Christianity and, 114
initiation to, 34, 36–37
Sedna and, 196–197, 203–204
Silatuniq, 33–34, 36, 113
sorcery and, 118
Sierra Club, 49, 89
Sila
changes in Inuit knowledge of, 111

changes in predictability, 22–23,
   185–186
combinations of, 29–31
description by shaman, 21–22
discussion of, 26–28
documentary on, 22
ethnographic conceptions of, 28–29
influence on pre-colonial Inuit, 154
meaning, 10, 19–20, 24–25, 27, 33
shamanism and, 36–37, 115
Silarjuaq, 29
Silatuniq
meaning, 29–31
shamanism, 33, 34, 35, 36, 37, 195
Western need for, 39–40, 213
Simon, Mary, 109
Simpson, Jeffrey, 201
Singer, Peter, 148–149
Smith, Adam, 122–123
Smith, Duane, 2, 5
social justice issues, 96–97
societal collapses, 164–165
sorcery, 116, 117–118, 129–130
Stephen, Michele, 129–130
Stotts, Jimmy, 223, 224
suicide among Inuit, 117
sustainable development, 84, 95
Suzuki, David, 63, 218, 219, 220, 222

T
taboo systems, Inuit, 194–199
Tainter, Joseph, 164, 165, 169
tar sands, Alberta's
destructiveness of, 162–163
economic benefits vs. northern
   warming, 98, 126
greenhouse gas emissions, 63
importance of, 132–133
northern oil development, 109–110
promotion of, 90, 91, 146–147,
   214–215
Tautu, Andre
explanation of IQ, 1, 22
polar bear behaviour, 8, 11, 188
pollution of the North, 110

This book is set in *Minion* and *Helvetica*. *Minion*, designed by Robert Slimbach in 1990, was inspired by the classical French and Italian typefaces of the late Renaissance period. *Helvetica* is a modernist, sans-serif typeface developed in 1957 by Swiss typeface designer Max Miedinger with Eduard Hoffmann.

Printed in the USA
CPSIA information can be obtained
at www.ICGtesting.com
JSHW081928301024
72691JS00002B/9